Technology
Touchpoints

Technology Touchpoints

Parenting in the Digital Dystopia

Loretta L. C. Brady, PhD, MAC

ROWMAN & LITTLEFIELD
Lanham • Boulder • New York • London

Published by Rowman & Littlefield
An imprint of The Rowman & Littlefield Publishing Group, Inc.
4501 Forbes Boulevard, Suite 200, Lanham, Maryland 20706
www.rowman.com

86-90 Paul Street, London EC2A 4NE, United Kingdom

British Library Cataloguing in Publication Information Available

Library of Congress Cataloging-in-Publication Data Available

ISBN 978-1-5381-6392-4 (paper)
ISBN 978-1-5381-6393-1 (electronic)

This book is dedicated to my husband, Brian, for all his loving repairs in our life. And to our children—Sergei, Zora, Langston, Willa, and Audre—for strength and compassion throughout the many ruptures of our daily lives. Your endurance and grace are a constant inspiration.

Contents

Acknowledgments

$\mathcal{I}$ began the proposal development process for this project in 2017, a time of storm and stress in my professional life and a time of transition in my research. These are the types of developmental disruptions that often drive change, and like many others, I turned to writing to make sense of them. There were many with whom I shared these musings, and I wish to begin by thanking colleagues at Saint Anselm College with whom I shared precursors and early versions of the project. In particular, I wish to thank Maria McKenna, Elizabeth Ossoff, Kathleen Flannery (Psychology), Jennifer Thorn (English), Kyle Hubbard (Philosophy), Dina Frutos-Bencze (Business), and Kelly Demers (Education), as well as former colleague Nicole Flores (Theology), now at the University of Virginia, and onetime collaborator R. Jamaal Downey (University of Florida). While never presented in whole or as part of this larger effort, the parts and pieces all benefited from their collegial attention and careful reflections of framing, approach, and areas worthy of further exploration. None of them are to blame for the many detours the project took, but I am grateful for their time and attention when these ideas were still nascent. I further wish to thank the Faculty Summer Research grant committee, which allowed me to apply their support to the completion of this project. Further, a special thank-you goes to friend and department Chair Paul Finn, whose stewardship of our department during times of his own storms and stress allowed me the administration free time necessary to bring this work to completion. Department administrative assistants Barbara Bartlett and Loraine Hobausz are also thanked for their humble support. Special thanks for the enduring friendship of administrative assistant and (JOA girl) Mary Molloy Rioux, who managed to manage me through life's ups and downs.

To the many students whose interests, questions, literature searches, and editing of earlier grant projects assisted in the development of this work, a personal thank-you. Especially helpful were students Sevda Islamova, Paul Viscione, Kelsie Cameron, Travis Zuchowski, and Jillian Rigby, who helped me prepare early drafts of materials for proposal development. Students Erica DeMatos and Brenna Leach were present, and contributed, once this project was officially underway. They demonstrated immense grace and flexibility as the moving parts of my life, and the project, came together. I never would have crossed the finish line without them or the able support of colleague Karen Van Wormer and psychology student Skylar Bottcher, who grew other areas of my research as I attended to this manuscript.

Thank you to my agent, Beth Marshea of Ladderbird Literary, and to my editor, Suzanne Stascak-Silva of Rowman & Littlefield, for seeing something in this proposal that promised more. Your faith, incredible patience with my pandemic delays, and confidence in the need for this work have made all that comes possible. Thank you, too, to the incredible Masheri Chappell of the New Hampshire Writer's Project, without whom I would not have had the courage to pursue the process of becoming an agented writer. Special thanks to S.D. Morehouse and Kira Morehouse, whose interest and encouragement kept me pursuing the project when I wasn't sure what it might become.

My writing, such as it is, has benefited from the collaborative writing of my early columnist days, as well as from competitions, retreats, and courses. I wish to thank Eric Ratinoff, with whom I once wrote a business diversity and inclusion column, and our editor for that column, Jeff Feingold of the *New Hampshire Business Review*. Their patience with my love of commas, long sentences, and curious questions have strengthened my voice without stifling its focus, and that is a rare gift for an unconfident writer to receive so freely early in her efforts. Truly these pages would not have been possible without their patience in 800-word increments. Similarly, Manchester Ink Link publisher and editor Carol Robidoux, who allowed me digital space on any range of topics and who made them punchier, more accessible, and more community-focused than I could alone, receives my deepest gratitude for her modeling of courage and creativity. Thank you to Joel Christian

Gill for motivation and for introductions to a community of editors that inspired the refinement of this project. I am grateful we met on that TEDx Amoskeag Millyard stage in 2014. I am so humbled by how far we have both come from there. Thank you to R. Jamaal Downey for the collaboration and patience in assisting me in finding a way to step into critical studies and public scholarship so deftly. Your work ethic and ethic of care is a rare gift in higher education and one you spend generously in support of the work. It matters that you do, and I am grateful to have learned from you.

The New England high-tech ecosystem and the community of technologists whom I have served and stewarded inclusion with are also thanked, especially Rica Elysee, Brenda Noiseux, Margaret Donnelly, Josh St. Cyr, Jennifer Gray, Julie McIntosh, Bethany Ross, Allison Grappone, Lauren Provost, and Elizabeth Hitchcock. Special thanks to all those who survived the "wives' dinners" on the climb.

A ResilienceCon writing retreat held in Portsmouth, New Hampshire, in 2018 and directed by Drs. Sherry Hamby and Victoria Banyard, and the community of resilience writers they formed and nurtured, is on my mind as this work concludes. Thank you to Brenda Tully and Karen Kalergis, two of the workshop participants whose work continues to inform and inspire my teaching and writing and who were compassionate and curious, two key ingredients essential to the writing life. Their attention and stewardship of interdisciplinary practice and accessible science has enabled me the professional permission to pursue the areas where my clinical and community work have focused. Other writers from whom my work benefited, either by convening the community of women of color writers, as Kira Jones did when she launched the Jack Jones Literary Arts fellowship for which I was a finalist, or as Lily Dancyger did, who conducted a creative nonfiction course in which I was able to participate during my 2019–2020 sabbatical year, are also warmly thanked. Making space for emerging writers is a mission for which few will offer praise, and yet it is that space that gives reluctant writers like me confidence to keep pushing through the doubts personal writing always brings. A special and profound thanks to Fordham classmate and psychologist writer Dr. Michael Alcée, who shared space and constructive support that made this work immeasurably better than I even imagined it could be. It's a cloudy gem, but one that would not be so without his careful cuts and polish. Thank you.

Friendships are the secret fuel of a writer's heart, and my life is blessed with these in many forms. Some friendships take the form of books. Carol Delle is thanked for always gifting the perfect volume at the needed time, and I so appreciate the care and time that took. Thank you to my collaborator and lifelong friend, Suzanne Delle, who never read a page but is somehow in every line by virtue of the calls, texts, and reassurance needed to weather the challenges described, and those left out. Maria Nalette is thanked for similar reasons. Your grounding friendship was a needed anchor when I needed it most. To my family of origin, including especially Elizabeth Evans, thank you for your education on the various ways local needs intersect with global realities. I may not understand your work and discipline fully, but I know enough to know my work would not be possible without our many conversations.

Finally, to my husband, Brian, and our children, who have always been the inspiration behind my fight to make the world a better place: thank you for the time and space that this project took from each of you, for the love that pushed it over the finish line, and for the anecdotes that made the hard part worth it in the end. I will never repay in time and space all that you sacrificed for this project, but I hope it makes you proud.

Introduction

$\mathcal{T}$his is a story about a family coming of age at the same time as smartphones and social media, a multiracial family coming into its own as windows into social injustice opened before our very screens, a multi-parent, multi-professional family with children living differently depending on which house and which combination of family members happened to be home. While it is a story about a family, it is really the story of the global actors, policymaking, the technology industry, and the impact between these entities and the social and global changes unfolding on our doorsteps. In the time since writing first began on this story, predictions and observations within the family are playing out in the broader culture, and the mirror the family story provides reveals a prism through which each of us, either as educators, parents, or policymakers, might come to understand the various technology touchpoints that impact, and can be impacted by, our individual and collective decisions regarding technology adoption.

Throughout the text, readers will recognize cultural touchpoints that reflect deeper human and technology development patterns, patterns we would all do well to understand, no matter whether or how we choose to engage in the ever-innovating digital frontiers. Informed by research and interviews with leaders in policy, human development, ethics, and technology, the book makes complex systemic challenges and findings related to technology and human development easier to understand and appreciate. Chapters 1 through 5 describe the incidents and research that help underscore the relationship between technology and social development. Chapter 6 outlines the key framework I and others propose to help support decisions around technology development and adoption. Chapters 7 through 12 help to outline specific sensitive periods in human development that are most likely influenced by technology adoption. Chapters 13 through 15 review ways to use the findings on technology's impact in guiding current adoption decisions and future developments, and the ways we might all urge policy changes that can improve our social and individual flourishing.

This work would not have been possible without the lived experience of being a parent with a technologist partner, nor would it have been possible without the opportunity to conduct my own research afforded by my career as a psychologist and a professor. Ultimately this work wouldn't have been written if these realities were not also coalescing with Steve Jobs's career and company. So, as we enter the family story, here's a shout-out to Apple and my early adopter husband; and to my kids, who challenge me with screentime limits and deal with my own personal limits of always assuming the worst outcome, and too often being in my head. Welcome to the chaos. The story is not a prescription, and the recommendations are not a finished map; instead it's a guide into perhaps one of the most essential conversations we need to navigate if we want to recognize what being human means in the digital world we have cocreated and that, increasingly, is creating us.

Parenting in the Age
of Anti-Social Media

My entire parenting life has unfolded during the growth of cloud computing: the release of Facebook (2006, oldest son and daughter), iPhones (2007, middle son), iPads (2010, middle daughter), and Influencers (2011, youngest daughter). As a psychology professor aware of childhood's most sensitive periods of neurological development, I ask myself questions I can't often answer:

1. How can I keep my children's cognitive development full and flexible and their reputations and images safe from outsiders?
2. Can I raise kids in this digital world who actually care about others?
3. Can I foster a techno-literacy that makes my children savvy to how others might manipulate their attention, take their money or their trust, and impart beliefs that aren't in line with theirs or our family's values?

It's one of the last lazy Saturday mornings before the semester begins. All five of my kids are sprawled out on the living room couch, each with a device in hand; iPhones, an iPad, and gaming controllers. They are all locked and loaded, yet nobody looks up as I enter.

"We're all playing Minecraft," says my oldest, well-practiced in underscoring the positives of gaming with siblings.

I couldn't sleep the night before, addled by anonymous notifications and emails listing my presence on a white nationalist database. My photo and contact information were posted in retaliation for a recent forum I had participated in on workforce diversity and immigrant economic integration. Before I knew it, the list of attendees, first published in a local conservative newspaper, was picked up by industrial-strength haters. From a few racially tinged headlines, the list became newsworthy, sparking increased vitriol in comment sections and in-boxes and precipitating the worst of the often-diluted racism and fear in our small-town New Hampshire community.

After four days, I was emotionally exhausted by my own rage over how the entire event had been run, and the resulting outcome. But by the weekend, I had finally formulated what I wanted to say to the reporters who had reached out for comment. I sat typing up my response while my kids played alone together in their own Minecraft world.

A robotic voice calls from the TV, where my kids are focused as they manipulate the console remotes. "Girl, you need Gram-mar-ly, lol, lol."

"Ahh!" my youngest screams, giggling at the online taunting they just received. They type back, and my son narrates actions, "The girl, outraged, types in reply. She goes all in on her troll." Laughter fills the room where they play.

I see cooperation. I see mental math skills as they calculate which resource to pay or pick up. I see typing and writing skills in the dialogue boxes they populate with one-liners and verbal "takedowns" and "trash talk." All is friendly, giggles from every corner. Big sister is providing overall coaching about the next part of the virtual world they will explore.

It reminds me of my research lab, and of my husband's high-tech offices. Individually focused effort but constant group feedback about essential tasks, unnecessary boasting, and purposeful focus. It's a stretch—maybe, but maybe it isn't. Maybe my kids, when they encounter trolls as I have this week, will be prepared, more prepared than I (sitting on my reply for four days) in navigating the virtual and literal worlds where such abuse might occur. Will my youngest know better than me how to detect the difference between friend and foe, intent and impact, wise and unwise thoughts to share, be freer to reply from a place of clear affirmation and support and bravery, where I hesitate and analyze? I don't know, but I see an ease in reply and witticisms that I suspect might be useful preparation for the tomorrow I can't foresee. Even so, there is an undertone of how wrong I might be. After I tell the kids they have fifteen minutes—my mom guilt catching up to me by lunchtime—I say it is time to play together in real life.

"This *is* real life," my youngest says. And I see she is not wrong, but I know she can't possibly be right. The hardest technology-and-kids question I find myself asking: How did we get here?

THERE ARE SO MANY OF US

Earlier, before the hate list, that same summer, we had met eight-year-old Ben on the beachside playground. A woman in her sixties was watching from a table as he swung his hands in the air and moved his knees sideways repeatedly. She studied the boy for a moment, then shouted, "Orange Justice!" correctly naming the dance performed by players of the latest online gaming trend, Fortnite. It had been popular in our area for a few months by this point and was about to celebrate its first-year anniversary.

"Hey, that's pretty good. You are up on the lingo." I laughed as I approached the picnic area where she was sitting.

I hauled over the bags and bottles from a day at the beach with five kids. She introduced herself as Ben's grandmother. She had sparkling, kind blue eyes, and her grandson was brown-haired, freckled, and a little chubby. While he was gleeful, she seemed worried. "Well, I don't know if I know it all yet." She sighed, shrugging her shoulders.

By now my two youngest, both around the same age as Ben, had joined in the hand shaking and knee sways. Ben changed up his movements, and my middle daughter shouted, "Jubilation!" She and her younger sibling joined in an altogether different dance move, this one involving rapid tiptoe steps with hands raised above their heads. "They play too?" his grandmother asked, her eyes meeting mine, hopeful but cautious. "Do you think it's okay? I mean, he seems like he's addicted to it."

I understand her concern. I shared it. Ben's grandmother and I had both been swept by a wave we hadn't known to watch for. The young children in our lives had become obsessed with an online, multiplayer, first-person shooter game that was marketed to entice our pop culture–enamored, athletic, competitive yet cooperative kids, who were content to take risks (in completely safe ways). While I was commiserating with her, I also felt obligated to demarcate the line between my younger children (then ages six and eight) and my older children (then ages ten and twelve). "Actually, *they* don't play yet, but their older brothers and sister do," I explained. "You are right, though. They do get caught up in their games. Those game makers know what they're doing."

Within a few moments, my little ones had begun playing with each other on the playground. My older kids arrived with drinks they bought from the concession stand. Sipping their drinks, they continued their conversation, which merged seamlessly into ours; it seems we both had been discussing Fortnite. Their conversation centered on characters, map locations, and weapon options. The woman's husband came over just as the older kids' talk turned to the weapons of Fortnite. Overhearing this, my face grew concerned.

"We don't allow weapons in our house," Ben's grandfather said pointedly, making eye contact. His grandson jumped in. "No, Pop! Just in the video game. Those ones." His grandfather sat down, certain of his house rules but not really following the conversation his grandson was having. He checked his phone, ignoring an incoming call and putting it on silent.

"We had to leave the house," the grandma continued as all the kids moved away from the picnic table to the swings. Grandpa looked back in our direction, "I don't know about this world he's growing up in. I sure hope they figure it out soon, for his sake."

It was a hot summer day nearing dinnertime. Her husband looked exhausted. I felt exhausted. A glance around the park revealed a mom pulling a wagon with a toddler holding a phone as a video played. The toddler's four-year-old sister was still on the playground, but she appeared to be the only one of the groups with much gas left. The sister was attempting to tell my younger children about her success climbing the playground equipment, but they were taking the opportunity, thanks to my own distracted chatting, to spirit away my iPhone to the playground. I watched them holding the phone in "selfie" mode and talking at the screen as they captured a video of themselves climbing on the playground equipment.

What were a middle-aged Afro-Latina psychologist and a couple of sixty-year-old white grandparents (who had raised four children to successful adulthood, plus six grandchildren) concerned about, anyway? Surely, there must have been something timeless about this conversation. While Fortnite was foreign and concerning for us two stably married and housed, well-educated women, with our typically behaving children, I sensed that a few years earlier we might have talked about YouTube. A few years before that, about tablets; a few years before that, about gaming consoles. And so on and so on, back to the 1970s, I might have overheard a woman like Ben's grandma asking *my* grandmother whether so much TV time was really a good idea. On the one hand, I felt calmly aware that we were having a timeless conversation. On the other hand, I had to admit to myself that I had already begun having doubts of my own. My older kids had experienced a school lockdown the school year prior, in the wake of the February 2017 Parkland, Florida, school shooting, initiated by rumors of an anonymous message seen on some other child's social media. My oldest son was receiving messages in the email account I monitored, inviting him to join a group playing games on Discord, a site that also was in the news for having prominent alt-right players.[1] Nikolas Cruz, the Parkland school shooter, was purported to be alt-right himself, having ties to white supremacists and reportedly known for hurling slurs at minorities.[2] I had had trouble getting another child to delete a profile on a social media site I had asked them to avoid; even my youngest children seemed to be too focused on screens at the expense of the imaginative play I had seen their older siblings engage in at the same ages only a few years before.

Mostly, though, I worry that I'm too late, or that my latest anxiety du jour will be supplanted by a new one, like a bad TikTok video replaying endlessly in my mind.

THE NEW NORMAL

While my family is larger than the current American average, among families with children aged eight and under, we are hardly alone. By 2016, the percent of children with access to some type of "smart" mobile device at home (e.g., smartphone, tablet) had jumped from half (52 percent) to three-quarters (75 percent) of all children in just two years. Seventy-two percent are playing games, watching videos, or using apps, up from 38 percent in 2011. This trend is not slowing or reversing. In fact, by my youngest's preschool years, more than 38 percent of children under two years old used a mobile device for media (compared to 10 percent in 2011).[3]

My family also isn't alone in the way our children's use of electronic screen devices has become harder to control. Even when families like mine try to stick to only selective applications and websites, there is a challenge in knowing what to trust and what to scrutinize, and plenty of bad actors looking to leverage that confusion. Apple settled a 2011 class action suit brought by families whose very young children ran up very large bills purchasing thousands of dollars' worth of digital items while using the iPad, and they promised to fix the issue.[4] While the solution helped, the actions the company elected to take were done not to protect users per se, but to ensure its users would still download those apps and allow their preliterate children to safely play with those devices. "Educational" apps—the number of which, as of January 2015, stood at eighty thousand in Apple's App Store—were and remain largely unregulated and untested.[5]

While we are not alone as a family, neither are we without guidance. Guidance comes in many forms. Recognizing the widespread use of digital media technology by young children, the National Association for the Education of Young Children (NAEYC) and the Fred Rogers Center for Early Learning and Children's Media issued a position statement on technology and interactive media in 2012.[6] At the core of the statement was the notion that digital technologies are another experience that can empower children's learning. Child-initiated, child-directed, teacher-supported, intentional play with digital devices can serve as a powerful and positive learning experience, and yet it can feel overwhelming trying to create such an approach in a classroom or home. Technology tools have clearly transformed how parents manage their daily lives and seek out entertainment, how teachers use materials in the classroom with young children to communicate with parents and families, and how these children learn.[7]

MANNERS AND MATTERING

On the drive home from the beachside playground, my boys are in the back, one reading a book and the other playing mental games in order to stay awake. He fails. The younger ones are riding in the middle seat. My oldest daughter wants to listen to music on her iPad (my easy listening station apparently pushing her patience to its limits). It's a device-free day for her, a rule I had set that morning, attempting effective mothering during a summer day that I regularly feel guilty for missing. I grant her access but immediately regret it. She pushes away the notifications of missed messages she received while we toured. She is moody after scanning them, and I watch her impatiently selecting different music. She shifts in her seat as I reach to pat her leg in reassurance and support. Seventh grade is on the horizon, and I recognize the signs of her emerging self. She settles into the ride home.

Once home, the younger kids immediately want my phone again. They set up a shot, narrate an intro, and advise their video viewers (consisting of them and me, since they do not have a YouTube channel) to "smash that 'like' button." *Who plays on a playground for an audience?* I wonder to myself as the earlier scenes unfold. What happens when playing on a playground becomes more about creating a video than it does about playing on a playground?

Questions I didn't think to ask myself when I saw them creating the video in the first place.

Why was that again?

In the background of the video, unnoticed by me during our time at the park, was the wagon-mom's preschooler, the small girl proudly narrating to my daughters how she had managed to scale the small climbing wall to get to the slide. My children had continued filming themselves, pausing briefly to acknowledge the slightly younger child and smile at her encouragingly. Just as quickly, my middle daughter turned back toward the camera and focused on the anticipated viewer beyond her hand. The little girl wandered off the video screen.

Revisiting the footage, my youngest children laugh and light up at their video. I smile too. It's charming to see the world as they saw it, the camera zooming by me talking on the bench, the action of the camera spinning around as each girl took turns filming and narrating their playground experience. They didn't have a YouTube channel of their own, but they had seen enough of other children's channels, and they had amateur and professional channels they subscribed to through my account. They had favorites, like CookieSwirlC, who started out making videos with Play-Doh and plastic eggs, and currently live-streams her adult self-playing Roblox, an online LEGO-like user-generated game world. Lately they are admiring a California woman, Sophie, who entertains viewers with challenges conducted by her contortionist self and her supportive, if less flexible, brother. While my children didn't have a channel, they certainly had a style, and it was one of enthusiasm.

Developmental researchers point out that three factors need to be taken into account when evaluating the appropriateness of a technology for a child: the content on the screen, the context in which the tech is being used, and the age and characteristics of each individual child.[8] Researchers such as Lisa Guernsey were interested in these "three Cs" as critical elements of children's digital media to support personalized or adaptive learning, as well as the ability of digital cameras and audio-recording applications to help children create their own stories and build narrative skills.[9] All the experts describing how to build skills in adults and children are revealing the one big message: These are no longer optional aspects of parenting. Teachers, school administrators, and other caregivers all face the challenges of preparing themselves, and children, for navigating their media-rich world in a landscape that allows for minimal supervision of children online.

PROTECTING, PREVENTING, PROCRASTINATING

The key to protecting children online comes first in understanding development across the lifespan, and second in understanding the important role adults have in sharing culture and values with young people. Whether you are sitting awake at night worried over who can talk to your kid in their latest Roblox adventure, or fearful that a child might be manipulated into an online suicide challenge, at the end of the day our strategies for protecting kids have to go deeper than simply setting a time limit on screen use or knowing our child's passwords.

What we know about all development over the course of a life is that knowledge comes before ability, which comes before sound judgment in applying knowledge and ability.[10] For example, consider a toddler learning to walk. The toddler learns how to stumble full-heartedly forward with great energy and, eventually, learns how to maintain balance while doing so: That's knowledge. The same toddler being able to consistently balance and ambulate in the direction in which it wishes to is ability, the ability to run and even jump while still maintaining equilibrium.

As caring adults, we understand that as that toddler increases the capacity to run and jump, stand, or balance, what's still lacking is the judgment of when exactly to use those skills, say, to stay safe on the sidewalk or to cross a busy road. It's why we run so quickly after a three-year-old who's approaching the edge of the sidewalk or driveway. Yes, the child can walk, but we don't expect the child to exercise good judgment or to cross highways safely. What does running after our children need to look like in a digital world, and what are the rules of the roadways we are protecting? Like Ben's grandfather, I find myself hoping we get this right before we've lost the chance.

Selfies, "Usies," and Attachment

$\mathscr{I}$n 1976, the year I was born, a developmental psychologist working with infants and parents created an innovative research technique using a brand-new technology: the video recorder. Dr. Tronick created films for use in his research on infant cognition. His actual research focused on the ways in which the mother and infant were or were not in sync with one another's attention and mood. In his lab, he filmed mothers' faces as they interacted with their infants, sitting in a reclining seat at eye level to the mom. He also filmed the faces of the infants, interacting and responding to their mom. He synchronized his recordings so that he and his research assistants could count and record the types of facial responses the infant made when interacting with the mother.

The films of these observational experiments became known as "Still Face" because Dr. Tronick[1] not only recorded what happened when mother and infant interacted, but he also recorded what happened when the mother stopped responding and engaging with the infant (who was secured in the seat and too young to physically maneuver for the mom's attention. Dr. Tronick speculated that infants could communicate even before they could produce language. What he captured is truly illuminating, even today, more than forty-five years later.

In his remake of the project in the early 2000s, a young white brunette mother makes joyous expressions to her six-month-old infant, who coos and gurgles in sweet reply. The screen shows mom on one side, baby on the other, and the viewer is able to see the beautiful ballet that happens between a preverbal child and parent. Mom's eyebrows shoot up; in milliseconds, her baby's do too. Mom's chin lowers; soon does baby's. Then a pivot, baby leads and mom is making baby gurgle sounds in reply. Then, a light flashes, and a subtle shift occurs. Mom looks away and then returns her gaze. Instead of a joyful, expressive face, she is vacant, neutral, and unengaged in her gaze. The baby notices right away. A frantic waving of arms follows his ignored cooing. Nothing from mom in reply. Leg kicks, then squeals, then a twisting torso shifting away from the whole visual field as he becomes so distressed that he almost tries to lean out of his seat to avoid even seeing mom's completely still face.[2]

They eventually called these patterns in the parent-child attachment dyad "ruptures and repairs," and these moments of vulnerability became known as "touchpoints."[3] The ideas revolutionized family support programs at every level of care, from case management, to healthcare, to infant mental health, an area of practice that was not as effective until the touchpoints revolution unfolded. The innovation beyond the filming and observing was in the approach to support families by offering guidance that anticipated the child's developmental milestones through age six but also recognized the likely regressions in other areas that would accompany that new skill, feeling, or idea. Baby slept through the night peacefully, but suddenly at seven months he's up all night again? Many families would throw in the sleep-training towel if that happened out of the blue, while Dr. Tronick and Dr. Brazelton were able to help the parent anticipate that change and reinforce the appropriateness of routine and responsiveness. Touchpoints (the center he eventually developed to train others in the approach, and the books he and Dr. Sparrow wrote by the same name) set off a revolution in working with children and their families, and the programs and books the center[4] created over the years guided millions of families through the hurdles of their child's early years, from infant sleeping, to toilet training, to social relationships for preschoolers. Many families drew on the foundation of that work to guide them through the tumult and transformations that adolescence brings, certainly another period of vulnerability in the life of the family and child, and when other family relationships and professional identities are likely also shifting.

BABIES, AND FACEBOOK, AND IPHONES, OH MY!

For many families, parenthood begins during a time when other developmental milestones are happening for the parents as well. Baby enters the world when mom is about to start a new management role, or baby enters the world when dad is about to join the military out of high school. Lots of adult developmental changes accompany and parallel the transition to parenting. For me, my parenting journey began as my professional journey was ramping up. As I welcomed my five children, I was also navigating a tenure-track faculty appointment, handling a professional credentialing process, and becoming an entrepreneur in order to build a clinical practice and revenue-flexibility during a fickle chapter of American Higher Education.

To say I had a lot going on would be an understatement. A large part of my ability to engage so many efforts had to do with my adoption and use of technology. In 2006, as a college educator in the greater Boston area, my university email enabled me to join Facebook. In that same year, my first child was born. By 2007, the iPhone had been released, and by the end of that year, my second child was born. Professionally, I used my phone to juggle the calendar, send documents to student researchers, and coordinate conference calls for community meetings. Juxtaposed in my personal life, I used it to capture small moments of nurturing my children, logging sleep schedules, and nursing schedules between courses. In January 2008 I posted my first pictures of my children on Facebook from my smartphone. My son is six weeks old in the photo, my daughter is twenty-three months, and I am thirty. My husband had taken the photo of the three of us playing on a slide near my sister's Florida apartment, where we visited her during my postpartum break. No selfie mode was easily available then; it's almost strange to realize that now, just fifteen years later.

Facebook was a place where I connected with alums, since for the first few years Facebook could only be accessed by those with college .edu email accounts. It became a place where I connected with others, like family and community friends. A few years later it became a place to share stories and advice. And all of that was a welcome thing for someone breastfeeding, bonding, and building a family and professional life.

Not long after I became a juggling mother, articles began circulating about the obsession with smartphones and their risk to kids. The headlines warned (mothers) about the risks of ignoring children in favor of screens; it argued for caution in posting information about children, and it sometimes even cautioned using the smartphone device around children, presumably because a mother's intent gaze at her engaged child was of special import to her even older and more independent child.

What was all the fuss about? Some, at least, was about mothering. Or rather, some was about the tradition of scoffing at women's adoption of technology and accusing it of denaturing womanhood. From brooms to vacuums, from stones to steam cleaners, technological advances often start in the home, helping alleviate some of the routine or labor-intensive domestic duties. In *Perfect Madness*, Judith Warner[5] has noted the ways in which technology that impacts the pace or efficiency of domestic chores associated with women (such as laundry, sweeping, child rearing) are often attacked in media as women's technology adoption rates increase. Warnings that women's time would be misspent if chores were easier and faster, were frequent. Eventually questions about authenticity and intention would creep into the headlines, accusing womanhood of engaging newfangled gadgets at the peril of their mothering and home-making. Maybe such warnings were useful; microwaved meals *don't* evoke the same emotional satisfaction as oven-baked goodness wafting through a kitchen might. Still, most households in the United States have a microwave that is used at least weekly.

And yet, tech is also a tool of motherly and fatherly attachment. The convergence of technological innovation in the form of computing, smartphones, and the evolution of social media occurred in 2007 with the release of the iPhone. A pocket computer with access to personal photos and online social media accounts amplified the spread of each technology. As the iPhone grew in sophistication, features and social platforms grew in reach and approach. Trends saw the shift of the microblog from 140- to 280-character limits, and the shift of the Facebook post from a soliloquy to Instagram images with compressed phrases instead of narrative. As if this range of personally stimulating material weren't enough, smartphones also held personal and professional email accounts, shared calendars for coordinating busy family schedules, and apps that tracked fertility or activity, that offered twenty-minute meals and capsule wardrobe organizers and secure encrypted files for work and personal business needs. In short, it was an executive dashboard that many moms and dads juggling work and family became more and more connected with.

But was it also an attachment disruption? More and more stories in newspapers, and then in newsfeeds, suggested that multitasking would interfere with efficiency and effectiveness in work settings. But for all the societal angst over parenting-while-smartphoning, I remember my own use was as a way to spend *more* time parenting, not less. It gave me a chance to attend to tasks that allowed me to physically hold and breastfeed my baby. It led me, through Facebook community groups or Twitter threads, to connect with other parents navigating their full lives. It introduced me to a term, *Attachment Parenting*, that suggested the most well-adjusted toddlers and academically successful kids were kids allowed to self-wean and stay physically close to their caregiver, even well into toddlerhood.[6] Cultural explorations suggested the North American tradition of separating an infant from its mother during transport, say in a stroller, was developmentally not as enriching as staying in close physical proximity to their caregiver, say through "baby-wearing" in a fabric wrap. Attachment was said to be the glue in human growth, laying the building blocks for the next developmental milestone in the child's life. Smartphones brought all those rich lessons forward, so I could push the stroller with good humor when that was the reality, and I could enjoy the wrap when the child and circumstances made that feel right.

In graduate school, my first research project examined the role of parent attachment in predicting children's later adult relationship styles. My mentor had examined how the impact of trauma, relationship quality, and child attachment styles impacted satisfaction in adult intimate relationships. There was lots of evidence to suggest there were links between these, and I followed the threads of attachment and trauma. As a new mom, I knew the power of attachment, and I was fascinated by the emerging parenting trend. Around the time of my third child's birth, there was a chance for me to train at the Touchpoints Centre in Boston. I was consulting on a project to develop a location for the Touchpoints family intervention model in my state and was sent to train in the approach as part of the "train-the-trainer" team that would ultimately bring the techniques and information to our state's social service system. We had invested the time and resources because the agency I worked with served a population deeply impacted by early and chronic traumatic stress. The stress had physical impacts on the clients, but it also had impacts on their emotional resources and parenting relationships. Supporting the clients in their attachment to their own children became a key clinical strategy as the agency supported more and more clients impacted by addiction and homelessness. We were drawn to the Touchpoints trainings because the information was based in human growth and neu-rodevelopment, because it provided guidance to families even if their child was developmentally delayed or atypical, and because it allowed for concrete suggestions and supports that were immediately repeatable, something unique in many multifactor family interventions.[7]

Over the many months of the project, we recognized that resilience was enhanced with academic success, which was linked to positive so-cial relationships within school, which was linked in turn to emotional regulation skills in young children. It turned out, as our advisory board dug further and further into the relationships between trauma and adult addiction, mental illness, and homelessness, that all of the many emo-tional challenges we sought to reverse and support in our clients were inextricably linked to attachment and parent responsiveness in infancy. We didn't pursue Touchpoints training because we intended to become regional experts in caregiver bonding, we were just trying to design a program for moms and their kids who needed a safe place to live while they tried to heal from their trauma. In trying to develop a flexible, low-tech, culturally and developmentally adapted approach to working with families, we came upon the Touchpoints approach.

The months following my second daughter's birth were a blur for me. I was four months' postpartum with my third child. I had just taken in a family member's child a few months before. Within six months our household had grown by double, I had sold and bought a home, and I had battled for tenure (ultimately winning). It had been a stressful preg-nancy to say the least. We had live-in help in the form of an au pair, a luxury of stable middle-income life that afforded enough income and home for a private bedroom and access to a vehicle. Even with all that stress and extra help, I remember the training, because in it I felt all the fears and concerns I had about my children and my fostered children were explained, and suddenly I felt the burdens of mothering lift.

The range of emotions, the depth of the connection, the power of the infant's responses were so mesmerizing, but also exhausting. This was neutralized for me in the training. Children have predictable mo-ments of developmental growth, and those disrupt the balance in the system between child and parent. Parent attentiveness and flexibility in the face of these disruptions return the balance, and the responsiveness of baby to the repair encourages the parent to continue responding flexibly. What a lovely way to think about my own poor moments as a mother, and to help structure natural supports for dads and moms in the program with far less help in their homes but just as much stress.

By the time of the Touchpoints training in 2010, I had seen how the iPhone became a tool of attachment as well. The camera in the phone became a routine feature by 2008. In 2010, when I came home from the drugstore with small photo books filled with photos printed from my iPhone, my husband commented on what a great idea it had been; by 2010 printed photos had already become a rarity. Even though in years to come, online printing companies would declare bankruptcy, I continued to flip through the photo albums when waiting for a fussy baby to settle. I loved flipping through them when I sat, back turned, in the doorway, trying to be present but unengaging as I tried to get three children under four to stay in their beds at bedtime. They were little peeks into the day or week. Eventually internet cloud computing meant that I could peek at my husband's photos as well, since both iPhones shared an iTunes account. When he traveled to bring internet intelligence to networked businesses, he could see the cute things I captured and I could see the cute models he was meeting at the industry trade shows he frequented. It worked.

For a while, our strategy in dealing with our children's Technology and Social Media (TSM)[8] access and use worked too. While the smartphone made connection with other parents and information possible, it also allowed us to encapsulate the blur of years that was unfolding. The device became a time capsule, and whether the content we captured was shared online with others or not, we became psychically attached to the device, and our children began to understand the nature of our connection with them through the window the device fills and the photos it provided. It was these little details that first drew the children into the device themselves. Seeing photos or videos of themselves taught them that the archive of THEM was inside THAT. Engaging with other media soon followed, and we soon realized proficiency and possessiveness became the marks of our children's attachments to our phones. As the devices grew smarter, their power to engage grew stronger along with ease of use. By the birth of my last baby in 2011, the iPad and the consumer product innovators they spurred released a way for iPads to be suspended overhead like a digital mobile. Stimulation was always available.

For iPads, photos and videos were not as immediately the draw, as the awkward and ungangly sizes and weights meant interacting in the same way as with a svelte smartphone wasn't possible. But what was possible was the viewing of photos and videos, and, soon enough, streaming shows and online games. The portal that was opened with the iPad was the customized viewing menus it enabled. Suddenly every member of the household could use the device to search and view the material they most wanted, in a screen size that felt like a personal TV. YouTube use increased dramatically, and by 2015 more than half of internet information searches were occurring within the YouTube platform. YouTube created an alternate, content-controlled, entrypoint called YouTube Kids in response to the realization that the main platform growth would not allow for monitoring as closely as most parents would require. In 2021, during testimony before a hearing along with other social media executives testifying before the Senate Commerce Committee on online safety for children,[9] YouTube announced its launch of "Supervised Experiences," a feature intentionally designed to provide greater parental monitoring.

Around the years between the launch of the iPad (with it enabling increased use of YouTube) and the addition of apps like Vimeo and more recently TikTok, the growth of the influencer became prominent. Media and ad buys began noticing that engaging with YouTube content creators could position their product or elevate their cause by funneling ad dollars to creators. YouTube leveraged this to incentivize popular content creators to produce more content through revenue sharing. The effect was thousands of creators of content across dozens of content and demographic niches. In her 2021 Senate committee hearing, Ms. Leslie Miller, vice president for public policy for YouTube, reported that their platform had compensated creators more than $33 billion in just the prior three years. Revenue sharing was a smart way YouTube leveraged its connections to its users, and soon the success of their approach created a host of new ways our family was connected to influencers.

It didn't start out that way, of course. In our family, the initial iPad, bought in 2011 by my husband for professional development, was used to access old episodes of *Sesame Street* and *Mister Roger's Neighborhood* clips. I loved these same videos as a child, feeling connected to Mr. Snuffleupagus and Lady Elaine. I recall these friendships and can sometimes bring back some emotionally rich memories of them that I held as a young child. It was a bit magical to be able to use the iPad as a time machine to share these relationships with my own kids.

The iPad also opened up some interesting ways to understand children's use of TSM. Its size made direct observation easier than the smaller devices did in typical child labs, like the one I used in the Media Engagement and Developmental Impact lab I codirected until my collaborator moved into other areas. In setting up our lab, we connected with Dr. Sandra Calvert at Georgetown University. We met years later when I visited her lab. She studies the types of emotional friendships I remember having as a child. Called parasocial relationships, her research documented that children (and adults) tend to form parasocial relationships with media figures and that these tend to last about eighteen months before a child moves toward another focus point.[10] Her work used naturalistic observation in children's homes, including bulky TVs being brought to kids' homes so the procedure could be standardized. The iPad became a way for lots of parasocial friendships to form in our family. *Influencers* soon became a word we used regularly as we tracked the variety of candy in Jillian's candy review, or the views of a sister and brother pair talking about their LEGO and toy sets. While the reviews were consumerist to be sure, what drew my kids back were the real children and families performing their everyday lives for them.

The research on the value and experience of parasocial relationships is compelling; these pseudo-attachments (since they are unidirectional, from the child to the target, and imagined) have benefits for children (and adults). They also have risks, and Calvert's research highlights the normative trajectory these relationships take when mediated in the form of children's TV characters. From this view, parasocial relationships are facilitated both by the medium of TV but also by the social and parenting environment. Availability of branded, character-based items like bedding, clothing, or toys can enhance the emotional connection a child experiences in parasocial relationships; parents not acquiring such items, or characters not being licensed for such purposes, will impact the degree of intensity the parasocial connection achieves. But as my children bypassed Dora and Boots (*Dora the Explorer* being a show my own younger siblings enjoyed while they were little), my kids dove instead into the YouTube-mediated world of influencers and reviewers.

While Calvert's early research uncovered the emotionally salient and personally meaningful connections young children make with produced content characters, others have examined and sought to understand how best to exploit the parasocial possibilities of user-generated content found on YouTube and now increasingly on other social plat-forms. The research of the ways in which interactive technology is used to engage parasocially with celebrities and leaders reveals that adults and children utilize these platforms and interactions in distinct ways from the TV character bonding described by Calvert and her colleagues. Instead, researchers identify ways in which the ability to create content deepens parasocial attachments to idealized figures, while also serving to constitute a digital narrative that brings personal meaning and identity to creators themselves with an inclusion of viewers. What is possible with YouTube (in a way that was less accessible in produced content chan-nels) was the ability to capture, replay, and revise one's sense of self and understanding of their place in their consumer and social worlds.

Predictably, this utility in the tools of TSM leads to lots of worry about narcissism and lack of empathy. If children are falling in love with themselves, the unsaid theory goes, they are probably going to fall out of love with everyone else. And as Calvert's, Brazelton's, and others' work reveals, humans benefit when they are in an attached relationship with others, whether those be social or parasocial relationships. What happens when the relationship is with oneself? What happens when the parasocial relationship is with someone who turns out to be harmful (as my children discovered when a favorite gamer was removed from a plat-form due to racist comments), or who turns out to have been harmed in creating the media for the parasocial relationship.

As we became smartphone parents, a lot of people were asking what it meant that so many mothers were online or how we might hope to keep our children safe online. Back in 2008, not many were asking what it would mean when children found their moms online. Or what might happen when other people saw your kids there. Or for some of us, what happened when kids found their sense of self there?

• 3 •

Memes, Meaning, and Me

Children are obsessed, parents are exhausted, grandparents are concerned, and this is how it probably always was. And yet...my time in tech-making spaces is filled with earnest and creative people, making the world better. In January 2019 at the Mills College Tech Intersections conference for BIPOC women and allies in tech, I meet two Mexican American twenty-year-olds from East LA who go by the moniker "Cybercode Twins." They are infectious, effective, and successful tech ambassadors. It's every parent's (and policymaker's) dream to hear their story and see their success. That's the inclusion possible in the space of TSM use and innovation. It's what draws parents in, the possibility of how TSM can ignite their kids. It also concerns parents in the way it completely captures their kid's attention and isolates them from any other humans in the same room.

I wonder what is taking away the building blocks of those great creative qualities I see in young TSM talent today? What if care, consideration, and creativity were altered by the use of TSM, and what if that alteration wasn't obvious right away? Is it too alarmist? Hasn't humanity's reaction to innovations of the past always been like this?

At twenty-two, I was preparing to change the world, in the way that earnest young women who have survived a lot often wish to do. I worked in a women's prison, helping bring professionals from the community into the prison to teach interviewing and budgeting skills. In retrospect, none of those things would actually help reduce their likelihood to reoffend. While the goal of the program I shaped might have been a miss, the work I did to connect incarcerated women to professional women was a hit. This connection impacted the professionals, as well as helping the inmates see themselves and their futures differently compared to how others might see their future as a result of their prison experience.

ME, MYSELF, AND I(NTERSUBJECTIVITY)

This magic—the magic of imagining a different future—is not to be underestimated. Researchers indicate again and again that seeing examples of people doing something can influence your own ability to do that thing, or some other equally compelling thing, for yourself. The area of psychology called social learning theory demonstrates how powerful this impacts us regularly. I knew that the women I was serving in prison were just regular women, women whose luck had run out, but who, for the most part, had been navigating the same risks and challenges that I had. I also knew the professional women I networked with had rates of poverty, childhood trauma, and exposure to domestic violence as well. While I got a lot of attention for connecting the groups, the reality was that they were not really different; both were navigating a man's world while using whatever resources and capabilities they could access. Sometimes those were socially sanctioned, and sometimes they were stigmatized, and stigma has a psychologically eroding effect on groups and on individuals who are associated or identify as part of that group. I knew they were just people wanting to feel prepared and purposeful in their lives. Many women have these same histories, but incarcerated women tend to have *most* of these events.[1]

A few years after I started volunteering in the prison, I was being feted in a Manhattan Harvard club dining room, having dinner with feminist writer Naomi Wolf and nine other young women who had done their own world changing. We were experiencing a weekend in NYC after being selected by *Glamour* magazine as that year's class of "Top Ten College Women." There had been a spread in the fall "back-to-school" issue, and our weekend in the Big Apple included makeovers and lots of serious luncheons.

A decade later, I returned to NYC to meet up with some of those women, and some of the five hundred other women who had been honored similarly by the magazine over the years. The women included Martha Stewart, Katiti Kironde, Sheryl Lee Ralph, and lots of political, scientific, literary, and business leaders. I embarked in 2007 on a study of these women, wanting to understand how being recognized for one's early contributions (those completed while still an undergraduate student) later impacted one's personal and professional interests. I wanted to know whether early life adversity had led to the activities for which the women had been recognized (some did; many did not), I wanted to know if the recognition had dampened or empowered their community impact (mostly empowered), and I wanted to know what they thought would have happened if they had not been featured—what outcomes did they feel would have been drastically different?

For most of the women, the magazine profile and the attention it directed toward them provided validation for activities that already mattered to them, and it gave them a boost that made them more willing to keep persevering in their next project. The media attention, particularly the attention they received back home, was also impactful for them. People who had fallen out of contact made efforts to reach out, and many of the women years later would beam with tender emotion over the memory of the attention and validation they received at a time in their lives when they otherwise felt uncertain and rudderless, typically during their early twenties and post-college years.[2]

What fascinated me about their stories was how common this pattern was for the women, even though the nature of their media recognition varied over the decades. In its first decade, women were honored for being "the best-dressed college girls"; in the second decade, they were honored as "the top ten college girls." It wasn't until well into its third decade that the magazine devoted to women's society and fashion labeled their contest "Top Ten College Women." Despite the fact that women were nominated first for beauty and fashion, and later for philanthropy, entrepreneurship, and innovation, the women described the experience of validation and how boosting confidence led the women to reach for new goals later in their lives. This is just one example of the power and influence the media has over us. Even while we are simply living our lives, doing the things we care to do, a blip of media coverage leads us to face ourselves and those who see themselves connected with us.

When I conducted those interviews with the prior "Glamour Girls," it was before iPhones became ubiquitous, and before fully curated information feeds were readily possible. It strikes me that the women honored between 2008 and now probably have had a very different experience with their honors, in large part because of TSM. What is probably the same is that each of us had to make our own kind of peace with the idea of being honored in a women's magazine.

For the women I interviewed, the impact of the award given at a point in time with primarily print-distributed media meant that they could choose whether and how to position the recognition they received. Would they list it on résumés or other professional biographies? How would they talk about it if it came up in an interview for a scholarly opportunity? Would the recognition by a women's magazine be received if they entered fields where men were more represented? Would revelations of the award undermine how seriously these women's contributions would be understood? For each of the women I interviewed, we understood that the source of the honor mattered, that it could influence how others viewed our other accomplishments and ambitions. Being celebrated as a woman isn't always appreciated in a man's world, as the prisoners and female business leaders of my college days could attest. We wrestled with what it meant to our future serious prospects if people learned of our award but did not understand what it represented.

Media exposure always leads to reflection on oneself, others, and the future in a way that few other experiences do. It is the bringing together the personal, the social, the anticipated, and all the emotions that it elicits. The women I interviewed about the award, and the angst some of them described, weren't odd or insecure; they were completely realistic about the impact of their image presented in ways they didn't control.

INTERSUBJECTIVITY

It should be no surprise, then, that over the course of early childhood, brain regions develop and rewire in response to aging, experience, and social interaction. One way we know ourselves is through knowing how others view and understand us. As we grow, we become more able to notice other people's emotional states, which is an important skill that helps us predict their expectations and responses.[3] This process is known as Theory of Mind. We cultivate this Theory of Mind across our lives, and it leads us to presume their beliefs (which may or may not be accurate). Ultimately it helps us predict their behavior because of the developmental process of intersubjectivity.

Intersubjectivity is the process of knowing others and ourselves, then knowing how to tell the difference between the two. And intersubjectivity has a lot more to do with how we behave around others than what we believe for ourselves. By a few hours/weeks old, infants can distinguish between self and other.[4] For a long time, pediatricians, mothers, and "experts" argued that babies were nonconscious until they became verbal. Fortunately, wisdom won the day, and people began really noticing the many ways that babies, even newborns only hours old, communicate and process information.

Today we know that newborns can distinguish between a touch on their bodies that is made with their own hand from a touch given by another's hand. Within only a few weeks, an infant can imitate a parent and can make a parent imitate them.[5] Those of us who have cared for infants know how intoxicating these pleasant early exchanges can be. At times, we can even see ourselves in the responsive child. What do we make of this power to differentiate between ourselves and other people?

It is an essential power, as it forms the basis for the continued differentiation we will make over our lives. It is the power that makes us, "us." Who we are is largely how we are, where we are, and what we are within a given social context. We will enter and exit various relational streams over our lives, but, ultimately, we, ourselves, are the connective thread over time and space, and that thread is formed by the intersubjective understanding of self that is created across these spaces. Knowing ourselves is a primary developmental task, one that we continuously revisit with each new experience and expectation we encounter. One way we make sense of our own intersubjective awareness is through creating representations of it—through writing, art, storytelling, and, ever more popularly, through self-creation and curation. Think of the middle school and high school bedroom walls and doors, and laptop covers of our youth, and you can see the way in which our intersubjective awareness gets displayed, practiced, discarded, and challenged through images, quotes, and figures. TSM provides ever more intimate opportunities for this display to unfold.

MOMS, MEMES, AND MESSAGE BOARDS

In recent years, there has been an explosion of media that enables quick creation and curation of images that convey social trends and understanding. In 2019 memes were thought of as bits of visual and narrative culture that transmit cultural information in microunits, although Richard Dawkins's 1976 text that introduced the term *memes* had a broader idea that memes were any small unit of culture, not just images and pithy jokes.[6] Memes that refer to a cultural moment and convey critique or commentary on the moment can be powerful as tools for subversive culture transmission.[7]

These funny little pictures offer visual jokes that poke fun of political, social, and even regional identities and understandings that are audience-specific. One group's memes are not understood or received in universal ways. Often those outside a cultural group won't be able to process and understand the intended commentary of a meme, as they are so bound in shared understanding and coded meaning. While some humor can be experienced across cultures, memes are most impactful for subcultures, especially ones in which a great power inequity is experienced.[8] Memes allow for the communication of frustration, invisibility, irony, righteous judgment, and general disdain that would otherwise be harshly and socially sanctioned if stated directly and without irony.

While memes may seem trivial, recently researchers identified specific functions and facts about their purpose and impact. Gal, Shifman, and Kampf wrote in their 2016 paper that

> Memes—both in the pre-digital and the digital age—are closely related to the process of norm formation. The memetic practice is not merely an expression of existing social cultural norms, it is also a social tool for negotiating them. The relationship between memes and norms is thus twofold: memes both reflect norms and constitute a central practice in their formation. In the latter sense, the process by which norms consolidate and/or change is characterized by a memetic-evolutionary pattern: the processes of adaptation and alteration performed by individuals may scale to mezzo or macro levels. A deviation from the existing norms may be rejected as a violation, reinforcing collective boundaries, or it may result in the subversion of these boundaries and the initiation of new norms.[9]

Transmitting and consuming memes has the power to shape our social identity, and in turn our intersubjective experience. They help us learn the social frames through which others (people whom we value or are curious about) understand the world. They become representations of entire knowledge bases, and they often use and transmit identity models to be replicated: That is, we can come to model our own identity after the one presented to us through media and memes. In the hands of me, feeding my TSM feeds in my early mothering days, my consumption and rebroadcasting of memes corresponds and documents my own cultural transformation of sorts.

Over the years of my early motherhood, I read and shared numerous articles about mothering and working, about decorating on a budget, about introducing technology to children. I was in short becoming a "mom"; or at least, shaping the version of "a mom" that I was mostly adhering to. I consumed versions of moms who made their children feel special and loved while also developing responsibility for their choices and a joy in serving others with their gifts. I didn't read many articles about how to raise a math genius; I didn't laugh at many stories about being a "soccer mom" or a "dance mom"; and my kids' activities were often pretty light, relative to others in our social circles. I was becoming the over-aware but laid-back mom, a meme unto herself; soon to be joined by blogging moms and entrepreneurial moms and overstressed and over-it-all moms.

After the death of Trayvon Martin, and then Tamir Rice, and eventually Michael Brown, I began following pages and accounts of mothers of Black and brown children, searching for ways to make my own anxiety about my multiracial children less intense, or at least focused effectively. Living in a state with such a low percentage of non-white residents, I felt isolated and alone in trying to navigate ways to advocate for my own and other children's safety. And the pattern repeated when, in the lead-up to and the aftermath of the 2016 US election, memes soon became ways to blow off steam or share "insider" jokes with those who feared for democracy and worried about social division. I can recognize in my review of timelines and feeds how I was becoming a mother, an advocate, and a citizen, engaging in civic and mostly civil discourse with people, and sometimes most inelegantly through memes. You can trace who you are today by reviewing who you've most recently been broadcasting yourself to be. You can shift who you become based on what you elect to consume and create. Media, and memes, have power.

The impact this had on me was not unique. The process through which memes become "me" has a name, used to describe the process by which we can at times adopt an identity as a whole unit, rather than as, say, an imprint on oneself. Interpellation happens when, through seeing an identity presented within our culture, we absorb the identity, and through this we understand ourselves through that model. The area of scholarship known as Critical Theory examines cultural identities transmitted through media and their impact on groups. This field essentially examines the levers and pulleys that influence what gets interpellated into what, and by whom. Our ability to recognize and tease apart the memetic contributions to identity does depend on our familiarity with creating and critiquing cultural content. Our skills in deconstructing our interpellation-derived identities are related to our creation of media, including drawings, videos, and fashion. Today, creation in the context of the TSM revolution is something that requires increasing technical access, while simultaneously needing less technical acumen to use effectively (an app for every purpose). This, combined with the laissez-faire rules around content as monitored by TSM creators and platforms, creates a potent mix of opportunities for images and digital media. And this powerful mix has the power to reveal us in ways we don't always appreciate, and to show us ourselves while also shaping that self that we show.

YOU'RE SO VAIN, YOU PROBABLY THINK THIS MEME IS ABOUT YOU

Ironically, it is this interaction between intersubjectivity and interpellation that makes us vulnerable to the nefarious TSM viewers of ours and others' self-presentations. The rabbit hole that social media feeds can become represents an understudied battleground, one where multiple cultural influences are being consumed but one in which subversive and harmful content can easily hide in plain sight. TSM provides new privacy concerns; experts were warning that easily located information about your interests, affiliations, and affirmations could be leveraged to steal your social identity, making future fraud against others in your circle easier to execute.

In 2017, the British government, in investigating whether and how pro-Brexit campaigns were targeted to certain social media users, uncovered the deep trove of personal preferences and information social media users of Facebook were allowed access to. The investigation revealed ways in which the user enjoying a specially programmed personality test ended up unsuspectingly sharing information about their contacts with the secretive company that paid for the game: Cambridge Analytica.[10] When the skimming of personal information was made public, Facebook vowed in congressional testimony given by founder Mark Zuckerberg that they would close the open door and remove apps that used similar techniques.[11] While the likes and preferences of your friends may not seem like particularly valuable information, it was precisely the type of info that could enable targeted ads—ads that might influence a vote or a major public decision. The Cambridge Analytica crisis often wasn't about you; it was about those to whom you were connected, what made them feel strongly about things, and how many others you could influence with your own posts and shares.

The nature of the data breach was such that while protocol might suggest remedying the user whose contact list was captured, it was those contact list members who suffered the impacts of the targeted ads. Especially if you represented the most active user of a platform or position, the amplified effect of your playing a personality test game may have been dramatic indeed. US Congressional hearings in 2017 revealed that similar ad targeting had influenced the 2016 election.[12]

This is how intersubjectivity is also leveraged within technology innovation and social media. By creating tech that allows for playful or thoughtful self-curation, there are embedded user motivators to revisit, refresh, and re-create one's self-presentation, drawing users in and creating self-presentation and cultural sense. Ads can influence behavior because we make meaning of our media, and we select behaviors based on what we perceive the range of available behavior choices is that we observe around us. The influence of one well-connected node of a social network could be effective because of the interplay between intersubjectivity, interpellation, and identity combined with the scale and speed of TSM. In this way, memes from one time become new material to comment with or communicate intergenerational meaning.

Meaning is the solder that binds attachment and identity[13]: We belong to someone and something(s), and we ride out the bumpy parts (or we don't) to the degree we ascribe these bonds as important and valuable. Psychologist Julian Rotter wrote about the formula that predicts behavior: What we value, we pursue; we pursue and persist to the degree that we value the expected outcome. It came to be known as expectancy theory.[14]

We can and do use this tendency for good; young people around the world earnestly raise funds for their summer church mission trip, and youth organizations connect kids to adults through shared athletic identity or community affiliation. Feeling connected to others leads us to want to support and help them, and, in turn, this leads to a reaffirmation of group belonging, thereby solidifying our identity as part of a group. When this goes well, it enhances our resilience, allows us to form and benefit from important social bonds, and fosters connections and good-will throughout our network.

When it goes poorly, however, it isolates us, exaggerates our sense of difference between ourselves and others, and leads us to engage in fear and scarcity thinking. In short, media and our meaning-making with it can set us up with a bowling league or with an extremist group; the process is actually pretty similar, even if the outcomes are so widely different. When that happens, our use of media and TSM to establish ourselves and our group's position can start to weaponize the very tools that we once relied on to locate community. How we experience this process has a lot to do with how we see ourselves connected, or disconnected, from the others in our social network, and as *Black Feminist Thought* writer Patricia Hill Collins points out in her ongoing work to expand the knowledge bases supportive of democracy, there are many whose interests are served by our disconnection.[15]

DIGITAL DIFFERENCES AND DISPUTED BOUNDARIES

I can remember trips in cars with my family as a young girl, driving by large apartment and office buildings, lights on in various windows twinkling over the skyline. It filled me with wonder to consider that in each lit window there was a person, doing something, and there were people in other windows, other buildings, other countries even, who were also doing something at that exact same moment. And, I knew, none of them knew what the other people were doing at that moment.

It is perhaps then no surprise that I became fascinated with Twitter a few years after its creation. Twitter was a platform that could allow one person in a room to simultaneously know what people in rooms far away were thinking or doing at that precise moment. But curiosity is one thing; knowing whether, and how, to make use of the technology is another. I was not a girl, but a licensed clinical psychologist, a professor, and a professional with many roles within my community. I had ethical codes and legal mandates to consider as I explored and considered where, if anywhere, Twitter might fit into my practice.

For us psychologists, Twitter represents a unique opportunity to think about behavior, professional ethics, and technology in new and interesting ways. It also represented potentially threatening and risky ways to present ourselves and to interact with the public. *Threatening* because a comment, repost of another comment, or favoriting a comment may position you in opposition to some professional, community, or personal standard. *Risky* because professional reputation, personal safety, and practice solvency may be impacted by use or interaction within Twitter.

Technology enhances our asynchronous interactions and facilitates verbal, written, or visual interaction and transmission. This often occurs in TSM without context and without current norms, and sometimes, at least for a time, with little to no consequences for those who interact inappropriately or intentionally cause harm through their TSM use.[16] The nature of our current TSM limits this to verbal and sometimes visual interactions, but does not yet enable motoric and somatic interaction in real time.

These sensory experiences when interacting with others provide valuable coordinating information, and this richness is lost in current limits of TSM. Furthermore, the extent to which these mediate our aggression and cooperation are beginning to be better understood as researchers tease out the ways in which our prosocial responses are shaped by our intersensory interactions and coordination. These are known to influence perceptivity, rhythmicity, and mimicry.[17] Which means misunderstandings and limited value meaning ("my peers will like me") may more easily be transferred than long-term value meaning ("my descendants will be grateful"). The psychological distance wrought by a keyboard and screen was made more intense by the insight and intimacy it also brought in its use.

On a family vacation in 2009, I set up my first Twitter profile. It took three weeks, in part because I couldn't decide on a username. I hold many roles as a professional, and I knew I wanted my profile to allow me to engage as a single person with multiple roles. *How best to message that?* I wondered. *Should I just use my name, a name that another public figure working in a mental health field already has and is well known for? Or do I choose a name that signifies a role?* I ultimately decided on @profbrady as my username. It represented a scholar on her way to tenure and someone committed to engaging all the work I do, as that of primarily being a professor. It also seemed to suggest that I took myself, as opinionated as I am, not so seriously. Within a few years, some of my business work meant that having more than one account would be wise, but frankly I've never managed multiple accounts very well. For a woman who spent her childhood moving through homes and different ways of communicating, I wanted my social media experience to be as coherent as possible. No easy feat when juggling so many online identities.

Twitter changed the world in many ways. Despite its own struggles to be profitable, the platform has ushered in an entirely different way of interacting with well-known leaders and celebrities, while also creating networks of like-minded but not real world–connected individuals. Twitter was where so many people were finding their circles, and by 2012, I realized that it was also a place where groups intent on disrupting a community could be effective and under the radar in their efforts.

Adria Richards, a Black woman working to boost company diversity for SendGrid, attended the PyCon conference. That day, after several demeaning and sexualized comments she heard and witnessed during the event, she tweeted out a picture to the PyCon organizing team indicating that two gentlemen seated near her listening to a presentation that included celebrating high school girls' engagement with STEM and tech, were audibly discussing sexualized jokes during the presenter's keynote. Richards knew the conference had a code-of-conduct policy, and her tweet was intended to alert conference organizers of the issue so that someone could come and address the issue with the men.

What followed was a nightmare of harassment for Richards. While the men ultimately were addressed at the conference, and their employers were equally alarmed by their actions leading in one case to termination, it was Richards who bore the backlash, not the men being professionally inappropriate during an adolescent's tech presentation. It was Richards, not the conference organizers, who suffered the online onslaught of abuse from men and from those who opposed any action to create spaces where women were not subjected to sexualized comments. And, as a Black woman, it was Richards who dealt with death and rape threats, who had to hand over her account to close friends so she would not have to deal with the abuse directly, who had to pay for therapy to deal with the PTSD to which the social media attacks, along with her company's decision to terminate *her* for speaking out, had contributed. It was Richards whose life was disrupted by the actions of sexist men. And as a woman who supported underrepresented men and women in tech, and who worked to help companies do better in diversifying their workforces, I watched and observed the ways the reactions isolated Richards. Eventually she would be vindicated as the broader cultural zeitgeist of #MeToo and #BelieveWomen unfolded,[18] but being a vocal, visible woman of color STEM advocate online with any kind of platform reach was, and is, a truly dangerous thing.

Within two years, the same pattern would repeat, this time for women gamers and writers who critiqued the misogynistic themes and art in video games. Women wanted games that reflected their realties, and if a new release did not do that, headlines would follow calling out this lagging edge of an otherwise-innovative industry. The headlines were meant to spur change, but the hashtags soon followed, and men's rights activists quickly swarmed the women writers and gamers who were using their Twitter profiles to speak out against the gaming misogyny. Gamergate, as this episode has come to be known, unrolled around 2014, and as it did, it revealed the interconnections between extremists, foreign actors, and homegrown trolls. Twitter, and the rest of the world, learned how particular users were easily targeted and harassed through the platform. Women, particularly women of color and often specifically Black women, created hashtags and "block parties" to help minimize the intrusions and alert others to the risks of engaging with certain followers. The effort taught authentic followers how to spot a bot, two years before the Russian campaign scandals were headlines.[19]

In some preliminary reports about Russian meddling in the US elections, these years—2012–2014—are cited as the years when Russian "troll farms" were learning and perfecting their use of social media as a cyberweapon.[20] In 2016, when then–Twitter CEO Jack Dorsey indicated the platform was taking new measures to better empower users to stop and report harassment, it was not lost on me or those women who had fought for years to get Twitter to deal with its very solvable issue of protecting users from coordinated harassment that action was finally taken when investigators started asking similar questions.

Reflecting on these early years of Twitter, I see the ways in which my community building in the virtual world impacted my community experience out in the nonvirtual world. By finding other women of color, particularly those who, like me, speak out about injustice and inaction, opened my mind and language to new possibilities, gave me clarity to refine my own voice, and connected me to resources and relationships that continue to shape me professionally today. Where my engagement with people better-positioned than I in my nonvirtual world was filled with protocol and expectation, my Twitter feed was a place where I could be an equal voice in important conversations, sharpening my own sense of purpose and self. Seeing other women changing their worlds made me feel like this was possible work.

What brought me a sense of belonging and purpose was also enabling isolation and anxiety in my emerging adolescents. The bonding my younger children had done with the candy and toy reviewers was beginning to turn into demands to consume what they were now selling, as licensed products with familiar faces began showing up at eye level in the cashier's line. On the one hand, kids navigate or avoid the threats of school shootings, and, on the other hand, being home on a tablet might make them depressed: TSM-facilitated interactions, confronting what that meant as parents, was something that mostly still felt impossible. But maybe this was all so much fretting; after all didn't our grandparents survive their bomb drills—mostly unscathed?

The Good, the Bad, and the Ugly

*A*t a dinner party in Oakland, California, I meet Lyon. In his late thirties, he is working for an online gaming platform my children use weekly. He reminds me of the energetic and charismatic church-planting evangelical ministers who sign their Twitter bios with "Church-planting Christ-follower & Dad." Except I know he isn't a dad, and his wife's professed Judaism suggests he may not be planting Christian churches around the Bay despite his joyful extroversion and his authoritative commentary on all things technology-related. He shares over many fine wines that he dropped out of high school, taught himself how to program, and climbed the ranks at every game company my children's fandom has made me familiar with. We pass around a bottle of hot sauce specially branded as a tech company giveaway, only mildly aware of its connection to our burning discussion about the values that tech companies spread themselves. Somehow this bottle of hot sauce is to remind us that data safety is a hot topic. Over dinner Lyon calls technology a "force multiplier," and it's clear from his elaboration, and the corny corporate swag gift, that forces can be good or bad.

Lyon in some ways represents some of the best that comes to mind when we think about the benefits of TSM as we know it. A hyperactive and bright kid in southern California, Lyon found school pretty boring, and by fifteen he was staying home teaching himself how to code using some library books he was delinquent in returning. When his junior year rolled around, it just didn't make a lot of sense to him to go back, so instead he followed a few muses and ended up living in the Bay area, where he landed job after job creating some of the foundations for the major gaming platforms we know today.

The story he tells about his latest company shares in some rosy features of the promise of TSM; a coworker who was a child when his company first started was apparently a huge fan, and eventually her tips and suggestions had such an impact that the game developers created an avatar based on her. Years later, she would join the company as an employee. What may have been a worrisome way to spend her leisure time as a child has grown into a vocation that her parents likely never imagined. Both he and his colleague's story reveal the great things about TSM: totally absorbing and malleable, in the hands of a bright and motivated person the basic tools and platforms can create self-directed creators who can impact their world in even more positive ways. With stories like this, it's easy to see why, even with some downsides, families overall are adopting TSM in greater and more immersive ways. If my kid's obsession is really just a way into a world that will work for them, then by all means! Take my money, my kid's attention, and leave me some insight about how to shape this for the best possible outcomes. Let the Lyons and the gamer girls of our world light our paths.

I thought about this on the flight back to Boston as I read Roger McNamee's[1] comments in the January 2019 *Time* magazine about globalization and technological change: Roads and platforms can carry good and carry bad, and they always have. McNamee is penning an essay to absolve himself of the negative effects he now realizes he and others unleashed in the form of Facebook, horrified by what he writes was an unmitigated and avoidable disaster in the form of the misinformation campaign impacting the 2016 campaign. Ben, the little boy my kids and I met at the beach at the end of the summer, celebrated the good things that came to him through his Fortnite obsession, joyously moving through the various dance moves he had picked up from watching his avatar celebrate their victories in the streaming game world. He and my kids were having fun, yet Ben's grandmother's concerns are easy to understand. Even while game developer Lyon touts the positive aspects of his journey from obsessed coding dropout to the leader of his new company, a headline at the end of last summer made me limit my daughter's access to his company's platform when one avatar used by a seven-year-old girl was apparently "gang raped" during free play when another game user created a glitch to allow him to hold down another user's avatar. Once informed, the company closed that game and barred that user. But the thought that a child without knowing what they were seeing could be exposed to sexual violence through gaming certainly led many parents to deactivate their game apps.

Our concerns, Ben's grandma and mine, come from places informed by headlines and heartache. There are no single answers when it comes to TSM, and while there are many great and several million harmless impacts, it is the vividness of the negatives that stays with us. We know that it isn't every TSM, we know it isn't every kid, or even every family. Yet we also know that at times our kid or our family or this specific TSM is not a neutral event. There are times when the impact is bad, not good. What Ben's grandma doesn't share right away when we meet at the beach is that last year Ben's parents had a hard time, a business went under, marital strife grew into evictions and separation, and now Ben and his siblings and mom are living with grandma and grandpa. Her concerns about his game play are in part due to her concerns that he isn't doing or saying things he needs to do or say because he's easily soothed with his game. And what if this isn't such a good thing?

Grandma has eyes on Ben in a way she never had before, and while she sees gaming helping contain an otherwise rambunctious boy, she also worries about what he isn't getting when he spends so many calm hours playing a game she couldn't easily join in even if she wished to. My newsfeed notification pops up as my flight ends. Two dramatic news stories of children hurt; one an eight-year-old girl who died after esophageal burns that happened after she daringly tried the "boiling water" challenge she had found or tried to make for YouTube. Another, the arrest of an adult accused of pedophilia and child pornography. He allegedly abused young adolescent children associated with a YouTube channel my daughter once watched regularly. Ben's grandma isn't behind the times, and she isn't naive; what she fears seems to be really happening online and in the real world. What Lyon says about technology as a force multiplier has me thinking about ways powerful and powerless people have navigated seismic cultural shifts, how the force-multiplying effects of past technology and entry points were navigated.

THE BAD OLD DAYS

Hanna Fry, the author of *Hello World*, a book about computing science's earliest days, reminds us that modern haranguing over technology's

impact on future generations and current culture is not a new phenomenon.[2] While I spend considerable time proving to my children's teachers their eyes rest on pages of books throughout the week, the truth is
they look at screens far more than pages. Yet if I was of a fortunate class
of women to have parented in the late 1700s after Gutenberg's press
had been in use, I would have bemoaned the number of printed pages
my children were consuming in a week, instead of participating in the
rhetorical and musical experiences they would otherwise be enjoying
outdoors. For many throughout human history, it seems that methods
to prevent the harms of technology have included forbidding its introduction altogether, as I might have done as a helicopter mother of the
eighteenth century.

Of course, ultimately, I would come to know what the parents
who have confiscated phones, and pagers before that, have come to
know: When it comes to preventing the adoption of a particular technology, we fool only ourselves. What we prevent is sought, and usually
attained much more easily than we considered possible. Blocking the
roads brings the good and the bad, and, in terms of McNamee's ideas
of technology access, only works for so long and usually that isn't very
long at all. When blocking the roads fails, or if we deem that approach
untenable for whatever myriad reasons, as the Pew Study for American
Life study reveals year after year, we seek to make the use as enriching
and as reluctant as possible.

Unless we have been presented with a narrative that TSM access
will be protective or promotive, if we perceive the device to be a safety
tool or if we perceive the game to be educational, we tend to open
wider the access points. And these access points then become portals
that we don't enter exactly the same way that our child does. This
leads to a distance between our experience of a medium and its media
and our children's experience and engagement with that same medium
and media. We may be in the same game world, even the same game,
yet our journey there and our experience within are not the same due
to our individual prior experience and exposure. It's been like this for
centuries, no matter the TSM innovation, but the impacts seem more
profound with the current and emerging TSM iterations I find myself
parenting around.

Even beyond these predictable differences between my experience
of a platform or product and my children's experiences with it is the

additional reality that much of our TSM experience is not simply mediated by user behavior, but it also is engineered through algorithms. We are directed to some options, not others.[3] We see some videos, but not (most) others. In short, when using similar platforms for similar amounts of time for similar reasons or ends, we experience very different TSM worlds while we are there. And being there in a way that allows for mutual viewing and understanding is difficult indeed. Ask any parent who has tried to reconstruct a child's browser history and you'll know how individualized the internet experience can be. So, what to do when access isn't possible to bar, and experience isn't easy to control or predict? What to do when there are lots of things going against your power as a caregiver? What cloak of protection are we left to offer as our children sail out on untested waters with uncertain vessels to contain them?

In the mid to late 1800s, the United States was experiencing its Gilded Age, as the standard of living had dramatically increased for those who had invested and innovated their way into the Industrial Revolution. Manufacturing and all manner of resource extraction and territory expansion was riling the imaginations of parochial village girls and boys. At the same time, thousands of people from Asia, Europe, and Canada were making their way to and through the New World, arriving in places where their hands would create, repair, or resell the items the growing wealthy classes demanded. While there was economic boom and cultural transformation, there was also great poverty, crowding, and despair within the poorest areas of many urban centers. In one, a pioneering social reformer named Jane Addams toiled to provide nutrition, healthcare, pastoral counseling, literacy and economics training, parenting classes, and direct financial aid to immigrant families in Philadelphia. The wealthy and well-educated Addams certainly understood her work at Hull House as being transformative for the immigrant and impoverished families she served, but she also understood that sharing her ideas and the plight of her clients was also essential and transformative work. If others were to benefit from the work she had begun, she would need to persuade an army of well-heeled and good-hearted helpers to follow her lead. She took it upon herself to write regularly about the lessons she was teaching and learning through her work with vulnerable communities. One particular essay she wrote at this time was entitled "The Devil Baby at Hull House."[4]

It's a tale that at first seems to suggest the superstitious naivete of immigrant women from Sweden and Ireland and Poland. In Addams's retelling of the story, she describes how although from different countries, women would tell similar stories. One story describes a young couple anxiously awaiting the arrival of a baby. The wife, humble and happy, seeks to make the best of their cramped and dingy quarters, salvaging whatever of the weekly pay she can through sharp stretching of the family larder. The husband, initially eager and excited, soon slips into drunkenness, so that by the time of the baby's birth, the fall has been so serious that the mother is left alone to fend for their wee one herself. Soon the mother learns that the baby is not what it seems; instead of a precious angel, she discovers their baby is possessed by a demon, and the battle for the very soul of their family is at stake.

In Addams's retelling, she reveals both the frequency and range of nationalities who would share this tale with her. Again and again she and her volunteers would unearth desperate women sharing their horrific stories. And she shares the impact these stories often had on the family; husbands chagrined in recognizing the effects of their own drinking stayed closer to home; daughters intent on taking up with one or another wayward street kid were halted in their tracks by the possibility of spiritual possession. In short, the stories, common yet unfounded as they were, held a power of protection for the pregnant women who were often otherwise powerless, especially in the context of being in a foreign land with conflicting cultural values. While they might not have had the language, the power—or the permission—to set strong boundaries for their husband's behavior or their children's relationships outside the home, they had the power and the permission to create an environment that would allow their children to experience grace. This power might not have been possible to demand, but with the cautionary tale of the devil baby present within many frames, the women had the power they needed to pull their families in closer at a time when drifting away was particularly perilous for them.

When roads and rivers brought good and bad in the form of new technology and new territory, it also brought stories. And while the devil baby was a story the travelers of the late 1800s told each other, I was struck by the stories moms I knew were telling themselves. Were the stories of bad outcomes from TSM used like the devil baby story, stories we told to try to control something we otherwise felt we had no control over?

OF MOMO, MISINFORMATION,
AND MOTHERS' TRUSTY WEAPONS

What makes the comforting steering of technology and TSM more effective is shared experience and exposure to culture. What makes it difficult to do is the dramatic proliferation and segmentation that has transformed the TSM landscape over the past decade. While cable certainly created divisions and segments, the territory covered by cable companies and the types of channel groupings they offered were only so varied. Essentially any community, or household, could rely on a few fairly universal ports for information. A shared reality (the intersubjective spaces overlapping) and a shared understanding (the interpellative environment having common foundations across groups) was formed through a slightly larger array of inputs. Yet the TSM landscape Ben's grandma was operating within included access points she could not fathom, never mind engage—from the gaming world Ben visited, to the Android app store and the iTunes app store and Google Play and Amazon Prime and Netflix and Twitter and Snapchat and Instagram and HBO Max and PBS KIDS and on and on and on. Similarly, Grandma's subscription to *Time* and *Education Quarterly* and AARP are likely far · removed from Ben's reach. Yet media streams do find each other, with replays of important game changes being discussed on the morning TV as *Good Morning America* plays in the background of a family morning routine, and dance moves from mom's own early teen years reappear in Fortnite avatar celebrations. All in all, however, the amount of shared media exposure has dropped, and expanded, dramatically with TSM innovation and adoption.[5]

What is lost when we lose shared viewing is the mediation portion of media engagement; interpreting and contextualizing the memetic transfer can only happen when intergenerational and intergroup viewing occurs. Providing meaning and understanding between groups and generations can only occur when a shared perception of something is formed.[6] Without this foundation, two TSM users see and respond differently from each other, and their bridge to shared understanding is compromised. Tensions that may come between the two due to differences in background can be amplified. In societies where there is economic and social insecurity, these different experiences can lead to fragile conflicts between groups. Whether and how we interpret TSM as positive, as a threat, or as nostalgic can rest entirely on whether we have shared understanding with others about the media content.[7]

In our current innovation cycle, TSM itself innovates and bifurcates so easily that wider distances grow between groups quite rapidly. For some users, culture trends promote a youth focus, so a bridge between their generational place and the ever-emerging youth culture continues, but this transmission is not universal given social, ecological, and economic migration trends. And even when many generations share in TSM use and creation, as they do on platforms like Facebook, there is still much strain between how they experience and understand the shared space and the information found there.[8] A post is never just a post; it is a shared communication style distributed across multiple niche channels with input relevant for only a select few. It's why even on my Facebook timeline, only a handful of my hundreds of online friends actually acknowledge any given post I might offer. What I say isn't for everyone, and not everyone will understand what I chose to share or post, nor would I for their posts.

Culture transmission requires shared experience; shared use is not shared knowledge. While younger people are faulted for relying on social media for news information, research investigating the impact of misinformation on sharing behavior revealed that older adults had greater difficulty accurately identifying false information from fact. The paradox that older users of Facebook were more susceptible to "fake news" than younger users strikes at our sense of things. Technology tends to lead to those sorts of flipped scripts; those younger seem to have a better handle on how to use the technology. Our cultural script for wisdom would have us believe that older adults are wiser in their use of social media; however, research suggests they may be more likely to misunderstand information presented through social media. TSM strikes at these very notions by demonstrating how vulnerable and gullible everyone is when it comes to TSM's magic.

Similarly, the risks and benefits of TSM are not even in their impact, given vulnerabilities and disparities that exist across the population. The individual and social group impacts are multiplicative and rippling in unknown ways. Gladwell wrote in *The Tipping Point*[9] how behaviors become a trend through the amplification of novelty. Those well connected to many others will do the best with crowdsourcing, those most visible in their use of a platform will generally suffer the worst from

Twitter mobs, but they can access protections more easily than those least represented. How that plays out is far from neutral or universal. And when precisely the harms of TSM are most likely or most severe is not such a shock either. To know the worst of what occurs, we only need to understand the stories that get shared. Whether the stories are true or not isn't even as important as understanding what they signal to us as we attempt to navigate across devices, platforms, and contacts.

In late spring 2017, I was fielding calls from our campus communication office seeking commentary on what appeared to be a rash of unfortunate suicides. In some ways I wondered whether the stories I was reading about the Blue Whale Challenge[10] were true; was it possible that more than one hundred people from across the world had been duped into committing suicide by an online "game"? The reports I found were from smaller local online news sites, some eventually getting aggregated and posted in larger outlets like Yahoo News. But the stories seemed credible, sad families unaware of their son or daughter's online participation in a challenge. Photos of them posting from eaves of roofs, artwork painted in blue tones, little breadcrumbs that strung together seemed like a suspicious string of maliciously orchestrated brainwashing. By the time I was reached for commentary[11] as a clinical psychologist, I was hearing reports that the initiator of the attacks had been arrested in Russia. But the horror only began again when someone supposedly took over the "challenge."

At the time I was asked to comment, the story seemed tragic and possible. Radicalization happened online all the time, sometimes directly, like through ISIS recruitment videos, and sometimes indirectly, as was reportedly the case with white supremacist terrorist Dylan Roof, whose trip through videos and chatrooms resulted in a pro-apartheid and KKK-style preference in programming.[12] It seemed possible that with enough time (more than fifty days) and enough strong engagement (which online interactions can make possible and surreptitious), many people could be convinced to engage in ever-riskier actions until they actually do cause their own deaths.

And yet, fast-forward two years and my youngest daughter and I are scrolling through game universes in the game world Lyon's company creates. A spooky photo of a birdlike woman's face, with large and exaggerated eyes, pale skin, and stringy, long black hair stares out from a thumbnail image for a new game world. My youngest daughter scrolls past it. "What was that?" I ask as she continues searching for the fashion competition world she is wanting to play. "What? Oh, that's just Momo. She's not for us." I ask how she knows about it, and she shrugs and continues scrolling. A few weeks later, moms on a Tech Parenting Facebook group post images of the same birdlike woman. This time the posts come with links to stories, some from two years ago, about an online suicide challenge involving "Momo," who embeds herself into seemingly children's video content only to coach them through how to harm themselves.[13]

It's a horrifying story. But at the time, I don't get contacted for comment. Within two more days, I see why. A story from 2016 apparently posted the "Momo" image, which was originally a special-effects sculpture. The image was taken when the sculpture was on display with a special-effects exhibit in Japan, and its intentionally scary visage had soon been associated with online urban legends about secret messages in YouTube content. By 2019, when "Momo" was popping up in my daughter's game world, she had already been exposed to the story on a YouTube channel that shared creepy fiction stories, but for moms like me with less experience tracking their children's emerging urban legends, her appearance reinforced the out-of-control stranger-danger feelings we feel allowing our children into online spaces. Just like the Blue Whale challenge, Momo didn't have to be real for us to take the reality of what she represented seriously.

And taking it seriously actually meant creating nightmares for ourselves and our children. When the real news story broke, the one about how easily false stories, like the Momo hoax, spread, many parents had already shown the image to their children, who in turn woke up with nightmares whenever the image would appear in their game or newsfeed worlds, which it did a lot as the viral news story got more and more airtime over a few days that January.

While I understood the Momo story was just like the story of "The Devil Baby at Hull House," I couldn't help but wonder why women today felt the same type of powerless fear the immigrant women in Jane Addams's Philadelphia had. What was going on, and what could parents do? How would it work when these were such good kids? In post after post on the "parenting with tech" Facebook pages I followed, I saw first-person stories, some anonymous but many not, about the impact their own or their children's TSM had been having on their family. Parents were concerned about predators using music performance social sites to lure unsuspecting children into real-world meetings. Moms shared gut-wrenching discoveries of their own or their daughter's friends sexting images showing up in password-required google drives. Dads shared posts about the dangers of Dank Memes and the chatrooms where boys and girls who just wanted to watch a gamer would find themselves exposed to neo-Nazi memes and pro-white nationalist rhetoric. Maybe Momo wasn't waiting in a video with instructions for how to kill oneself, but there sure were plenty of other life-threatening ways TSM could be coming for us or our kids.

And it is our kids who appear to be most impacted by the TSM wave we are currently riding. Throughout the past decade on college campuses, there has been an increasing awareness that mental health services are not optional programs, but are essential, particularly as more and more students with previously less treatable conditions receive help and are able to maintain their course of studies, even during the psychiatrically vulnerable time of emerging adulthood. And yet often the uptick in needs for mental health services are cited as being due to the treatment advances that have allowed more and more students with anxiety and depression able to complete their studies. In fact, the incidence rate of mental health issues in college students isn't just about treatment advances. It seems that TSM use and proliferation has also had a major impact on student well-being. And the latest research makes plain it isn't only youth's mental well-being that suffers.[14]

THE UGLY TRUTH ABOUT FAKE MEDIA

In 2009, I was up in the wee hours of the morning on my way to the bathroom while pregnant with our third child. My husband wasn't in bed, and a trip downstairs revealed him, phone and laptop in hand, looking worried. Within a few hours, the news would report that Twitter had been hacked, exposing some key leaders from the company to social engineering. Even a company on the rise was vulnerable to attacks.

As news accounts of the harms of bullying, the lingering effects of online threats, the live-streaming of school shootings, the accessing of porn, and the boom in online challenges and youth risk-taking capture the parental-attention microscope, we hear less and less about the true vulnerability TSM creates. The vulnerability of isolated individuals with a limited ability to discern real from fake, relevant from irrelevant, daring from devastating. We rightfully worry about those with bad intentions luring our children into poor decisions and unsafe places. We spend less time than we should, however, worrying about the various ways known threats can be leveraged to create even more realistic and treacherous TSM, and the ways this itself can get reified and amplified due to tech innovation and policy stemming from machine-learning efforts. And the most treacherous advancement that has occurred for TSM in some ways is the one thing that seems impossible to control; the live-streaming of events has opened brand-new portals into darkness that current platform-based regulatory approaches seem impervious to stopping.

Lyon is right: Technology is a force multiplier. Much good is amplified through TSM adoption and innovation. Our dinner host that evening, Julia, created coding opportunities for children half a world away; boys failing class can find a way to challenge and support themselves; and people who might not otherwise meet can find each other and create things that didn't exist before, all without even leaving their homes. The good: TSM opens worlds and gives us a future we can't even imagine. Of course, if that was the only story, this book would be a lot shorter. TSM also causes some real harm. Overuse can lead children to be overstimulated, tired, and disconnected from their families and the activities that bring them pleasure. Adults can and do find ways to use TSM for all sorts of damaging outcomes. TSM provides doors and roller coasters even for those who don't know how to open only safe doors or ride only inspected roller coasters. The bad: Ben and some other kids seem vulnerable. They don't seem like they know how to set limits, and there seems to be lots of intentional and unintentional malevolent actors they seem ill prepared to avoid. The fact is, some kids and adults are more vulnerable to the bad, and it isn't always who and how we think, as the election misinformation and white nationalist incitement reveals.

The ugly in this story is that the mechanisms humans have typically relied upon to mitigate the seismic shifts that accompany TSM in every era—strategies like shared mediated experience, slow introduction into larger and larger groups, and wise counsel over when and how to best make use—have been thrown overboard by the very nature of the current TSM innovation cycle. Our best strategy—to train situational awareness and judgment through social modeling and mediation—is difficult, if not impossible, to do due to algorithms, device differences, and platform opacity. We, parents and children included, are alone in navigating a world on which we have increasingly less influence as nine companies exert control over billions of users (humans) worldwide.[15]

Just like the mothers visiting Hull House, we are left using the classic art of culturally transmitted intergenerational protection known as storytelling. "The Devil Baby at Hull House," after all, is a cautionary tale for dads not to drink away the family purse, a reminder to children that goodness and evil are present from birth so minding your p's and q's is proof you aren't a devil after all—no matter how the streets treat you—and above all, young women make sure the love you have is fierce enough to endure even the hardship of a devil, as you'll have no more control over any of it than you would an actual demon. Time to adjust your expectations.

If this isn't the basic advice that parents and children are left with, I can't imagine a more fitting allegory than to keep everyone close so the devil doesn't have a chance to snatch your happiness. Since we can't control when devil babies will appear, we are left only with doing our best to understand how they are likely to be spawned. In Chapter 5, we will explore the times when humans are most sensitive and vulnerable to TSM, and when the risks of TSM going wrong are likely amplified.

Sensitive Periods, Disparate Impact, and Sensitization

*K*eeping loved ones close, communicating, and developing shared understanding and meaning: This is the stuff of life. Mothering looks a lot like this when done well. The repetition of this over time is indeed the key to our human evolutionary advantage, the development of our "cognitive niche."[1] These behaviors (proximity, communication, creating shared mental models) are the foundation where human attachment occurs. Attachment is the developmental task of infancy, and one that continues across our life course, with specific developmental moments where attachments are influential in orienting human behavior toward evolutionarily and developmentally important opportunities, such as learning adult roles, selecting community, and developing intimate relationships. Scientific interest in the power and importance of human development as a result of attachments formed in infancy was of interest to John Bowlby, who would teach Mary Ainsworth and who would create the theory that now animates so much of our understanding of healthy human adaptation.[2]

Yet Bowlby was not the first to be fascinated by infant and caregiver interactions. Others before him, like Charles Darwin, wrote one of the first scientific papers on infant development,[3] though he is known now mostly for his Galapagos Island travels and *Origin of Species* opus.[4] Still, between Darwin and Bowlby, scientists have appreciated some special significance between infants (of many species) and the caregiving behaviors observed between them and their parents. While Bowlby and Ainsworth established the enduring significance of early attachments with parents,[5] the questions of whether there are specific moments of attachment opportunity, and whether these are associated with specific developmental tasks, began to increasingly become the focus of development specialists working with children much older than Darwin's infant, Bowlby's mothers, or Ainsworth's toddlers.

ORIENTING SENSITIVE PERIODS

While human babies would continue to enchant and engage scientists, it was ducklings that began the search to determine if human attachment was tied to specific periods in development. When Lorenz termed "imprinting" to describe the attachment behaviors newly hatched ducklings exhibited toward the first fowl or mammal they encountered after hatching, it was soon applied to human infants.[6] It shouldn't have been; imprinting and human attachment are indeed different things, serving different developmental needs and with different impacts and implications. Human infant attachment to a parent after birth is not what a baby duckling is doing when it imprints.

The idea, however, that there were times when some impacts might be more significant than at other times was a reasonable and important idea, and there *are* sensitive periods of human development. What is less understood, however, is what implication there should be when this is affirmed in science. The sense that bonding might be tied to critical developmental windows referred to as sensitive time periods captured the imagination, and methods, of developmental psychology, for much of the mid-twentieth century.

Along with ducklings and their quest for maternal care, there were kittens wearing goggles painted with horizontal or vertical lines.[7] In studies seeking to understand stereoscopic vision, and attempting to model the neural and optic changes associated with deprivation of experience during critical periods of optic nerve development, the kittens' visual systems were manipulated to create environments by painting horizontal or vertical lines on the goggles, depriving the kittens of access to horizontal or vertical planes in their depth perception development. Kittens displayed agnosia for vertical or horizontal edges in their environment depending on the type of visual deprivation to which they had been subjected, impacts that affected the mobility and sociability of the kittens as they grew. There were impacts on adult cat functioning as a result of visual deprivation experienced during their kitten-hoods. Would such impacts be observable in human children? Ethically depriving human infants of experiences should not happen, but natural observations of childhood deprivation do occur; one such observation occurred during a time when America was reckoning with its relationship to two socially transformative technologies: television and space exploration. Our connection to the impact of screens on earth's children (or at least North America's children) begins as we seek disconnection from being bound to one planet.

SENSES AND THEIR LIMITATIONS

Children left to fend for themselves in nature is a perennial theme in fairy tales and modern children's literature. In history several actual children who survived childhood by being protected by animal packs, or who survived severe deprivation, are known. In 1971 it was baby Genie who captured the attention of clinicians, and through their media-captured narratives, engaged American imagination.[8] Raised by severely mentally ill parents, Genie was left mostly unattended in a crib for several years, from eighteen months to age twelve or thirteen, when her horrific living conditions were discovered and she was rescued from her parents' sparse home. A question had been burning linguists and cognitive psychologists for decades, as to whether and when a sensitive period of language development might exist. Research with multilingual individuals, deaf and hearing individuals, and others seemed to suggest such important realities; however, naturalistic studies were not possible. Baby Genie changed that, and as her language and relationship skills progressed, the capacities of human language acquisition began to be revealed. A year after her rescue, reports on the impact of television on children were released; Genie's sensitive period in childhood marked a sensitive period in the childhoods of American children, as the nation's leading developmental specialists considered the long-term impact of televised programming exposure for children. The report had been years in the making.

In my own home, I begin early on to see the influence of TSM on language. My children's awareness of phonemes wasn't only from hearing me read Robert McCloskey books like *Make Way for Ducklings*,[9] but from seeing me type words and hearing me narrate their pronunciation as I typed into my smartphone, something I was doing a fair amount of in well-child visit waiting rooms in 2010 and 2011. My children, who were then infants, preschool, and kindergarten-aged, would accompany me on the infant visits, and their waiting-room antsy-ness would be handled with a guided game of Angry Birds. With my children calm in the presence of these games, I was able to text baby weigh-in updates with Daddy.

Time and again, our animal models of human developmental sensitive periods fail to align with human developmental trajectories. The evolutionary cognitive niche we occupy as a species continues to thwart our efforts at reductionist explanations for individual difference or performance. Our ability to compensate, rehabilitate, and generate in response to stimulation and enrichment is very high as humans, and this makes it harder to disentangle one system from another when trying to single out systems sensitive to certain experiences across time. What is sometimes easier for us to measure, though certainly not easier to establish intellectual agreement around, is when there are different impacts for similar events occurring at developmental stages.

CAPITAL GOODS, DISCERNING DIFFERENCE

Two Januarys, twenty years apart, mark sensitive periods in technology development, both of which were televised into the homes, and awareness, of America's children at specifically sensitive times in education policy debates. The events befalling the launches of both the 1967 *Apollo* and the 1986 *Challenger* shuttles[10] represent sensitive periods for both children and the capital goods those children would shape: the public transparency in improving space safety in the *Apollo* investigation, and the public relations via education approach to STEM engagement envisioned in the *Challenger*. During a visit to a graduate school friend's house one January in 2012, I sat in a formal sitting room in New York. My grad friend, whose family life began before our PhDs were minted, sat next to her then-ten-year-old daughter. The girl was typing furiously with her thumbs, having been working to coordinate a visit that afternoon with a friend across town. Every few minutes, while her mom and I caught up about graduate mentors and classmates now directing hospital wards or earning large grants, she looked up and reported to her mom snippets of the text conversation.

My friend's daughter was demonstrating a mastery of conversation and logistics, something I tuned in to because my own texting was woefully unmatched to the dexterity and fluidity I was witnessing. For this slight but connected child, hopping on and off a text thread with friends felt as native as shouting upstairs to a sibling in her home; a natural way to communicate. For me, text messages felt like work email, and I wasn't ever a fan. But the tasks of a ten-year-old female are not the same developmental tasks as an adult female with offspring. We had different work, so texts meant different things to each of us. But for both of us, our ability to text, finding ourselves in that room, was connected to the January socially sensitive periods that had occurred to our space and defense programs just decades before.

On a cold January day in 1986, my classmates and I kneeled on the Berber carpet squares of our suburban elementary school and watched as nine astronauts, including a teacher from our state, would perish in the *Challenger* explosion.[11] On that day, our elementary school suffered its first national tragedy; we would hear the sixth grade teacher from down our hallway weep as she realized how close she had come to dying, having been one of the teacher alternates participating in NASA's carefully planned public relations campaign. What had been a moment of local pride and distinction instantly became a day that marked our community forever. More than a decade later, I would attend graduate school in large part because the legacy of one of the other astronauts, Ronald E. McNair, would pave the way for underrepresented college students to enter academia and become the most diverse faculty body created in the United States.[12] A desire for public support of NASA's expanding budget led to an increase in STEM education emphasis. It also led to the tragedy; in the pressure to produce a launch that aligned with the predetermined schedule, warnings and safety checks were ignored or overridden in what has become a case study in every General Psychology text teaching about groupthink.

Certainly, as a McNair scholar who benefits from the work Ronald's brother Carl and family continue to do in his name, work that elevates first-generation scientists who otherwise would have few paths into rigorous careers that require a doctorate in order to obtain an entry-level role, I feel like I was set into a stratosphere I could not have attained without the guidance and support provided at a pivotal vocational time. But more than simply feeling connected to an astronaut's legacy, or in my case, several astronauts' legacies, I recognize now how very much today's TSM issues, constituencies, proclivities, and challenges are related to our species', and our country's, desire to detach from earth and reach space; that while I might feel attachment is a key to navigating TSM, its very existence is the outcome of a desire to be detached from the very physical, and human, planet earth.

All our innovation, inquiry, and impact has been, and will continue to be, shaped by the mid-1900s leaders' conceptions of space, desire for fulfillment of the continued Manifest Destiny doctrine of the preceding era, and fueled by imbalanced economic growth in societies shaped for extraction rather than enrichment. Telecom, satellites, video transmission, portable personal communication, graphene, virtual reality, cryptocurrency—all are tied to space investment and governmental demand, often itself tied to corporate interests as evidence from campaign finance contributions; indeed, these interests seem consistently invested in administrations, since resources tied to these efforts create markets for innovation, commercialization, and monopolization. While such a defense ministry or department-driven cycle of technical innovation is not new in human history—we had chariots, and now we have armored cars, of course—it is significant that families in 2022 are navigating their landscape in the context of an innovation cycle shaped largely by the ethos and concerns of the United States and England in the 1950s and by the markets and fixations of a few figures in the United States, Russia, and China in the 1980s and 1990s, but such it is.[13] The fact that it is also has bearing on how we might think about the social technology touchpoints we navigate today, and the ones our children are likely to navigate with their children is relevant to any discernment we might engage around technology and parenting.

In considering impacts, sensitive periods may not be relevant only in individual or family cycles, but also in culture cycles influencing the context a child navigates. For some cultures, there are sensitive periods when the influences and impacts of TSM are particularly fraught or fretted over. Every era and environment will have its own concerns, and each TSM will engender its own contexts, shaping the span of interactions possible and unfolding probable paths of influence. Philosophers considering how technology represents human evolution and innovation, call any innovation, including a thought rendered as code, a capital good. The purpose of a capital good is to generate excess production, not just for a given manufacturer as we might consider in an economic model, but in this case also to generate excess production for humans, as a group, to hold knowledge and capacity without having to store it within any one individual. The inefficiency of human knowledge acquisition, behavioral expertise, and network transmission results in collaborative efforts toward innovations that reduce the effort and increase the efficiency of these capacities, so that focus on other areas of challenge and opportunity can be mutually pursued.[14]

I am considering this memory of my friend's daughter and the astronaut relevance of January in my life, when I connect in 2021 with Lisa Guernsey of *New Republic*.[15] I am interviewing her as the originator of a framework that helped parents and preschool educators think about what a particular TSM might mean for their families. She contributed this work shortly after that grad school friend visit, when the adoption of iPads into children's play and learning spaces was increasing and researchers wanted to maximize the use of this powerful tool, in a way that wouldn't rob them of other important classroom experiences. Her framework to help parents and early educators consider digital adoption—Child, Context, Content—would become an enduring way that families, and eventually TSM creators themselves, would consider when designing and adopting TSM for children and classrooms. There were forums created; Lisa mentions one in our Zoom interview, a conference called Dust or Magic, where interdisciplinary experts would unite to consider ways to improve the impact of apps for children.[16] This group, combining developmental researchers, children's educators, and app developers, meets annually. Together they work through how to scaffold a child as young as aged three on equipment capable of communicating with NASA astronauts orbiting the Space Station.

When Lisa and I meet, she is still humbled by the endurance of the work she did then as a fellow at *New Republic*, framing the conversation about tech with families. Her work now is shifting, looking at the gatekeeping to information represented by Jim Crow practices and laws in the United States throughout the school integration era of the 1940s and 1970s.[17] The 1986 *Challenger* explosion enters our discussion; even there, in seeking to understand the impact of TSM on youth, we are confronted with the legacy of one lost that day, Ronald McNair, who had uneven opportunities to engage his Southern library due to Jim Crow.[18] It seems even in considering modern stories of equity and historical portals for accessing and developing information literacy, we can't escape talking about space exploration—and barricades to knowledge themselves constructed by policies about access to information, particularly for children, and especially for Black and poor children, being disparately controlled by adults. Capital goods are not evenly spread, but their availability is impactful for all in significant ways.

These are not the typical concerns of developmental psychology; instead, the field tends to train its lens on the biological and ecological phases and changes. The desire to know *now* what will be *then* is one that psychologists in particular have cultivated methods and measures around for more than a century. It is perhaps one reason the psychology discipline has been charged with being out of touch, missing patterns or profound social influence because it finds itself myopically focused on distinguishing difference and predicting similarity rather than shaping the conditions that generate the foundation for difference or sameness.

Psychology, both the field and the subject, has long been absorbed with comparing individuals and circumstances to better understand outcomes. This child in this home with *these* sets of experiences can have strikingly similar outcomes to this *other* child, with other sets of experiences, who somehow also has the same success or disorder noted in the first child. Similarities despite differences are the bread and butter of psychological science. When we find different outcomes with the same starting information, or similar outcomes with strikingly different starting points, psychological science has some work to do, and it takes this work seriously.

While psychology has long been focused on differences and similarities, the most central work currently within the field is appreciating disparities, when group differences are reliably predicted and explained by resource and access differences rather than by individual behavior or character. Disparate impact is what we seek to measure when assessing the effects of an event that do not impact evenly across populations or groups. In the work I have engaged over time, comparisons between zip codes in a community often reveal large differences in lifespan, quality of lived years, and other health indicators. It was this evidence of health disparity that was instrumental in highlighting the effects COVID-19 was having across different economic and ethnic groups in the United States. It is this reality, that well-being is not evenly distributed across communities, that is rarely discussed when exploring the implications and impacts of TSM adoption in homes and communities, despite being named as part of a national strategy to improve US competitiveness in science and technology in 1997, under President Clinton's Council on Science and Technology.[19] The impacts of TSM are most definitely unevenly experienced, and therefore so are its impacts. But what might those be, and what can we do, even if we are to know, for our specific family or communities?

The concerns parents express regarding TSM and the disparate impact it may have for their children center on the ways in which TSM seems to monopolize their attention, how that attention becomes monetized through the in-tech behavior tracking, and how monolithic the industry influence on their children seems to be; as well as what happens if the push for STEM education, and the TSM adoption it has helped justify, leads to the monolithic pursuit of tech to the detriment of life, love, and human interaction. Who will we be if such dominating occurs, and how can it be helped when so many forces are pushing in these directions? Who will our children be? Who will we be?

SENSITIZING SELF

Who we are in response to what we experience is part of the human condition; we are impacted by what we ourselves impact. And what humans create has an impact on other humans, the earth, and as increasingly, our galaxy. As humans, our sensitivity to other humans has benefits and purpose. Sensitivity to others around us, and our sensitivity to their experience, assists in helping humans form bonds and, through those bonds, achieve important tasks that allow for social group advancement. Indeed, that is the primary social function of infant attachment, forming the foundation for social attunement that will enable successful development. Whether parents know the mechanisms of action or not, however, many rightly question the influence that TSM might have on their families and its ability to attune sensitively to what really matters, rather than simply what is attention-grabbing. That TSM creates distortion fields impacting a given person's approach toward others is hardly news, and the concern that TSM use with children might have particularly sensitizing effects is also not new to the modern digital media era of TSM; rather, the impact on emotion and self-perception was even of governmental concern to those navigating film media with children in the early 1900s, and again in the 1930s' rise of radio programs, fraught for children due to their "emotion-arousing" impacts.[20]

As I gather the research and family observations that help me crystalize this pattern of disparate impacts in TSM adoption for young families, the reality and extension of these patterns begin playing out globally and nationally. The area researchers concern themselves, regarding sensitive periods for or disparate impacts of TSM adoption, also includes considerations around how the media might sensitize those who engage with it.[21] What tech was adopted, who adopted it, and where they used it ended up predicting what information would be promoted through TSM platforms such as Facebook and Google. The power to use one's TSM affinity and proclivity to influence social behaviors was beta-tested as isolationist and xenophobic messages received boosting and network spread to predicted and stunning effects. This itself increased TSM engagement, ensuring a distorted relationship between aggrievement-fueled political messages and TSM corporate success. And the cycle fueled additional harm, in the bystanders whose access to images and videos of other humans being harmed as a result of social oppression, unrest, or opposition was suddenly increased, while their access to community care, validation, or support was dramatically reduced. Polarized messages resulted in polarized access to support as escalating disinformation circulated, leading into and out of the information cycles of 2016.

Attachment, of our attention to reaching the moon, to the crime storyline in a radio broadcast, to the action of dynamic first-person shooter video games, or to interacting with others in Facebook groups, always involves emotion. It's something Dr. Alexander Kriss, director of the Fordham University Community Clinic and writer of *The Gaming Mind*,[22] speaks to in clinical work with patients. Games are vehicles for meaning, goals, rewarded effort, and connection: Patients derive and convey their self in sharing game attachments. The same is true for all TSM; how we engage and where we attach are meaningful to, and aligned with, the shaping of our psychology and social worlds. The research conclusion is that even the same event can have impacts that differ, depending on timing, intensity, frequency, experience, and response, and that for some, these impacts will be long-lasting and permanent. Sensitive time points, in a child, family unit, or social group, can all profoundly influence the impact an event has on an individual. The kinds of events made possible because of TSM are not neutral and may not be temporary.

Sensitization from TSM engagement has been documented and is implicated in youth mental health impacts, in the rate of racial trauma reported in Black TSM users, and in survivors of state violence. This type of sensitization occurs when observers identify with the individual or the group who is being harmed in the image or video. Another type of sensitization that occurs as a result of TSM is one that is often explicitly sought when used educationally; as a topic is presented repeatedly, an individual's ability to learn and master material about that topic increases, since they have become sensitized to what is relevant and irrelevant about a topic, and so are able to identify relevant information and learnings more easily and efficiently than those who are naïve to the topic. Sensitization can also refer to the impact that prior exposure has on a person's ability to recognize current salient information; as Hong Kong protestors adopted a hand-salute gesture from a fictional American dystopian fiction series, Chinese officials who were not sensitized to the implications of the symbol were stymied in their attempt to control the TSM messages spread through their otherwise tightly controlled internet sphere.[23] And no issue has been longer studied, with such little results, than the impact of violence in video games as a desensitizing effect leading to psychopathic deviance otherwise absent from our social landscape. The sensitivities to the power that TSM has on our sensitivities has no bounds.

As a new mom, and a newly minted assistant professor leading into the inauguration of our country's first multiracial president, with a wife who was the first in her family to obtain a law degree, and a multigenerational, multicultural family anchored in every corner of the globe, I was sensitized to attend to the policymaking and parenting that was unfolding as the Obamas became America's First Family. I was anxious and pleased for my own family to see a Black family, an immigrant and multigenerational family, in the White House. I was pleased to see a working mother, a college professor—things I identified as and with— as a leader in government and service. I appreciated that reflection, the stories that were generated in news and magazine lifestyle sections with such profiles being advantaged, and the opportunity to bolster values important to our home through their example: gardening, family bowling nights, *Hamilton* remixes, and the like, from what I now see as a rather naïve period, but one that does have deep and warm memories in our family life.

Psychologist Maureen Craig and Jennifer Richeson, among others' work, suggest that political headlines signaling racial shifts, or corporate messages signaling bias, are all detected and determinative of individual and group behavior.[24] Racial salience adds power to messages, and behavior as a result is always altered, but not always in ways that individuals can identify directly. What I experienced in those years as we made family decisions was rather idyllic in terms of TSM exposure, but I was, by the beginning of their second run as First Family, starting to notice. Some of this was in my noticing myself as a mother of color, as a mother of Black children, of a Black son, and as the result of living in a state where both are so limited in numbers, they become memes and headlines whenever the minorized population is reported on nationally—like the headline that garnered the white nationalist hate list maker that led me to worry about how to respond one July morning in my kitchen, months after the Obamas were no longer the First Family.

There are no periods in human development that are *not* sensitive to the impacts of social interactions. In even micro-exchanges, humans respond to social interaction, actual or anticipated; remembered or ritualized, we respond to others. We might not always be as sensitive to the humans around us as when we have just welcomed a new member of our family, or experienced a personal accomplishment, but we are always attuned cellularly to our social worlds. Knowing this can make our strategies around TSM somewhat more grounded, if not always able to be well-informed. When are we sensitive to the impacts TSM might have on our sense of self, possibility, or potential? Who might be most at risk for the negative effects of TSM or when might others be most at risk? If we can't know everything about what can go wrong, can we know enough about how things can go wrong that we use this to guide our plans? Is it limiting us socially to consider yet to be in TSM based on what has been?

THE ATTACHMENT THAT MEDIATES SENSITIVITY

In the fall of 2021, as my youngest child turns ten, and only one day before the start of my children's fifth-, sixth-, eighth-, and tenth-grade years, I can't say whether the model of TSM engagement we provided has prepared, harmed, helped, or hampered my children or family. I can say I know the degradation of distraction, preoccupation, anxiety, disorganization, and distance that comes from the busyness that TSM facilitates so effectively. I can say that these things, in the context of crisis or conditions with deprivation, are a unique risk and vulnerability profile that bears attention, by parents certainly, but also by providers, educators, and family support workers.

That is, until they can't be, and when this happens, the conflicts that impede innovation will direct the possibility of the pathways that are open once the conflict passes. These are not always predictable. The impact of a given market player within TSM life cycles and human lifetimes can end up having outsize impact on entire social groups. We are all astronauts because of the sensitization we experienced socially as our nation committed itself to manifesting humans' long-hoped-for defeat of gravity, atmosphere, and jurisdictions. Ideals of patriotism that united a country through impoverishment, and through global strife, intentionally shifted their unity message away from the earth (which was riddled with environmental ruin, along with civil and human-rights conflicts) to space—something beyond the mere human squabbles and resources of earth, to something metaphysical, and in doing so resulted in increasingly complex metacognitive skills to manage and control as it expands.

The mediating of the media (no matter the medium) is the behavior and relational condition that fosters secure, organized, and prosocial engagement, with TSM but more basically with other humans. We can contribute to these types of patterns by using the tools psychotherapists use in their work supporting infant mental health, the work that Dr. Tronick and Dr. Brazelton, mentioned in the last chapter, had also pioneered. Observing behavior—nonverbal behavior and verbal behavior—leaving room for the user to imagine the more collaborative responses and providing opportunities for that behavior to be practiced and affirmed, is critical. Doing that in the relational spaces we share with those we love is important, and to the extent we can facilitate environments where these can be supported, we can enhance rather than erode our relational and intellectual repertoires, even when TSM appears isolating. In Chapter 6 we will explore the time periods when human developmental potential and tasks are most aligned with TSM impacts. Technology touchpoints are the times when parents and policymakers need to tune in to ways they can limit specific risks associated with specific types of uses and communities.

· 6 ·

Technology Touchpoints

$\mathcal{L}$ike most parents, we found ourselves overwhelmed with navigating discordant messages about young children's developmental needs and their screen time. Recommendations that children have no screen time before twenty-four months[1] compete with ads for educational apps that promise shape recognition by age two.[2] Stories about the advantages to be gained by educating through online or app-delivered programs like Bedtime Math or ABC Mouse proliferate.[3] We and other parents are increasingly inundated with media content that is touted as educational and with messages that early engagement with media improves our child's school readiness and knowledge. At the same time, we are admonished to monitor our child's technology access.

It is this paradox that I am most asked about when brought in as a speaker in high schools on how to avoid creating a zombie "screenager." Both sets of messages seem important. I, like many parents, worry about getting it right. A memory of the "many under five" era of my family life is the evening my husband and I took three Pre-K children and a newborn to see the 2010 documentary film *Babies* playing at the only independent theater near our tiny city. Only one other couple was in the theater, and they just so happened to choose the seats in front of us. I am not sure if they got more chuckles from watching the adorable antics of Mari, or from hearing my son shout/whisper, "She's feeding her with her boodles like you do, Mommy!" in response to seeing baby Penelope nursing. Frankly I can't decide which of those made me laugh more, either.

The film was never part of the Touchpoints trainings that Dr. Brazelton and others developed over the years since he and Dr. Sparrow first published their volumes, still resources for new parents today. Even so, it maps beautifully to the developmental milestones that they articulated and laid out with families in their newborns through Pre-K–aged children. In *Babies* we see the period when newborns' visual systems develop acuity (four months) become a time when mother and child negotiate more during feedings and when a child is at risk of being weaned if nursed, or started prematurely on solid foods if bottle-fed. Then we see crawling and sleep disruption begin to occur (seven months), and the times when infants can orient to their parents' social perspectives (nine months).[4] Each normal human developmental progression is seen in each infant, no matter their cultural context.

It is a wonderful depiction of the main tenets of Touchpoints, which is that human is human; the rate and specific timing of these milestones may vary due to individual and contextual factors, but the processes, and the developmental needs met through them, is normatively human and predictable. It can also be useful in guiding parents through periods when attachment and the infant/caregiver relationships are likely to be most tried, most sensitive to rupture, and most in need of repair or attention to reorient a bonded attachment system into functional dynamics. In other words, babies may not all walk by twelve months, but all families with twelve-month-olds will have walking as a developmental milestone with meaning, and they will derive and place meaning on whether the milestone was met or not in ways that conform to their expectations.

Likewise, my observation of the impact and milestones facilitated and foiled by TSM leads me to reflect on the early childhood touchpoints Brazelton and others outlined and to consider what developmental milestones across post-infancy might look like. I also consider what technology-specific implications there might be in the intersection of these developmental milestones and the impact of TSM within the context of those milestones being experienced, navigated, or captured, and what is the needed and useful anticipatory guidance that might be imagined as possible if we were to observe, narrate, anticipate, and support ourselves and families in ways similar to Dr. Brazelton and his many trainees.

I believe that focusing on the attachment system, its risk of rupture and its opportunity for repair, is the focus worth having as a parent in navigating technology use and policies in our families, and I suspect, based on a few years of ethics stewardship and exposure to feminist care ethics, that there is a true social value in thinking similarly in our policymaking toward TSM, big tech, and innovation cycles. The pandemic, along with the global climate and geopolitical changes that it has wrought, also make me attuned to the variety of alternate social systems where TSM policy might be considered in ways that yield very different stances to TSM than our recent frameworks have provided.

Taken together, there is a great deal of attention in my ideas that focuses on the conditions for positive, prosocial, and sustainable community living. It's perhaps not the most economically, technically, or even politically astute modeling of the scope of the challenges associated with TSM, nor would it be the most historically informed or accurate. Each of these disciplines is outside my own areas of training. The only thing I have studied well is trauma and resilience, and, because of that, I know I have an informed perspective on attachment and repair. It is this limited frame that shapes these ideas, but as my internet connection continues to bring me into the treatment rooms and testimony chambers on cases concerning victims impacted by TSM, I am more and more convinced there are technology touchpoints that we can anticipate, consider, and repair with support, insight, and guidance.

Here's my attempt to do that, with full concession that my own choices in navigating TSM, for myself or my family, certainly haven't been well-guided by much. So, perhaps it's unfair, this many pages into our story, to suggest once again this isn't a recipe for success. I do feel compelled to offer guidance while acknowledging how far from faithful to these insights I myself might be, from one TSM or parenting decision to the next.

Even at five years old, my child understood what a QR code was. For most parents, our conversations about media—about what's real or not real; what's true or not true; what we should question or accept—become fewer and fewer the older a child gets. But if we take as true the fact that knowledge comes before ability, which comes before judgment, as the touchpoints model articulates so elegantly, we begin to understand why these value-specific, frequently reiterated, and present kinds of conversations are increasingly important the older our children get. This is because they know how to get new information, and they have the ability—at even very young ages—to do that, but they still don't know that it is right and good to believe, or if it is right and good to defend, or what is important to disclose or keep safe.

The media literacy education we need is one that not only teaches parents how to use tools to promote online safety, but that teaches parents how to have conversations throughout their lifespan about core values and beliefs: how to have hope when things seem impossible, how to practice gentle forgiveness and mercy toward others, how to practice humility and curiosity in ourselves. It is a media literacy education that shows children how to use tools as tools, that teaches boundaries and etiquette and civility, that teaches how to create as much as it models how to consume, that is explicit in its value structure. And perhaps most challenging in the context of data that travels across the globe and is stored, we need opportunities for children and youth who are learning these lessons to make mistakes that won't forever harm themselves or others.

When I feel most anxious about these matters, I try to take a deep breath and think about what we want our children to know about their value in the world, what they need to understand if they ever have to walk anything back or tell you something that you might otherwise really not want to hear. I think about what I need our children to know about where else they can turn if they ever felt like they needed to talk to someone about something they're going through. I think about helping them understand who safe adults are and how to access them if they need to. What I'm taking from the touchpoints in thinking about my own children's TSM engagement is an appreciation for the stable pattern, if not the timing, of knowledge, ability, and judgment. The value of having a sense of the technology touchpoints might be in providing the way we can bring the consciousness of this pattern into our own choices in adopting TSM. The concept also challenges us to consider that the arguments and battles we have had or will have over the specific place and timing for a given TSM will shift, but that the primary goal in our parenting and policymaking is to attend to the needs in the parent-and-child relationship that are manifested by the TSM battle we are then navigating. The battles, the TSM itself, and the parent and child's relational system needs will be in flux across a lifetime, but what won't be changed is the repair to the relational rupture and the benefits that attending to that relational need will have on the individual child with TSM, and the family system navigating the needs it has for connection.

While Brazelton and his collaborators explored the developmental milestones of children through age six, my work with patient communities and our own household shows me that developmental needs and milestones occur across one's life, and necessarily then our engagement and experience with TSM will shift based on these changing capacities and needs. Here I will outline briefly what developmental opportunities are broadly aligned with shifts in TSM use and impacts. Between infancy and the decline of older age, humans seek connection to others, to a sense of purpose and place, and to express and create in their unique and valued ways. How TSM facilities these needs depends on the life-stage and developmental objectives of the family system. From bonding, to bridging, TSM allows humans to attach, and detach, and therefore our use of it will benefit from appreciating fully the power and function of attachment.

BONDING

For all families, TSM decisions, relative to children, begin before a child is delivered in whichever regional birthing situation is likeliest. A mother's monitoring of her menstrual cycle, the timing of an Uber call relative to a delivery order, even the types of coupons issued via an app have all been TSM-related indications of pregnancy that have impacted parental attachment.

Unsuspecting partners learn of their impending arrival through auto-deliveries and sponsored ads before seeing their partner's glow themselves. And the tracking and data extraction only quickens from there, with every new app downloaded, every new Facebook group joined, and every new Zillow search executed. Deciding whether and when to obtain a cell phone for your middle schooler is hardly the first TSM decision a family makes. Prenatal ultrasound screenshots,[5] complete with estimated gestational age and location of facility, are regularly tweeted or posted as personal revelations that a couple is expecting. Family genetic profiles, along with photos of 23andMe genetic testing, are regularly posted with enthusiasm, revealing matrilineal DNA codes that are unchanged across female descendants and shared in an era when we have every reason to believe data will exist forever in ways recallable for generations, times when capabilities beyond our imagination would be possible and in governance circumstances that are unpredictable.

What are the bonding-related risks and TSM relevant guidance worth considering with new families and very young children?

Most will concern parent adoption behaviors. There will be different family needs that will drive their relationship with TSM. Was the birth predictable, did the mother feel supported, was the baby's arrival as imagined? These indicators suggest that TSM use will be related to schedules, communication with family, and documentation (photos, videos). There might be differences of opinion across families in how or whether things should be shared on social media. There are likely different uses and preferences in TSM use between new parents, and guidance affirming the value of documenting and taking breaks is worthwhile.

For some families, however, the arrival is not neutral or positive, but rather traumatic. For these families, TSM is likely used to assist in information gathering, processing, documenting, ruminating, self-soothing, or avoidance. Use will likely focus less on recent relationships, and more on external activities or inaccessible relationships. TSM use may interfere with the available attachment systems and in turn can result in earlier and more pervasive adoption of TSM in infancy, when pediatricians and developmentalists all assert TSM use poses the least value and most threat developmentally relative to cognitive structures primed for enrichment and pruning.[6]

Following the shift in focus to academics, the end of summertime is always tricky in my household. Planning an early-school-year birthday around new-school class schedules has made it so that over the years the end-of-summer party photos featuring that school year's core friend group ends up being driven more by our household's older children and their friend groups rather than it does by the birthday child. For every child's first birthday, however, there is always a photo book filled with images of the friends and families who populated our days in those times. Facebook reminds me of this when I don't think to look at the physical birthday books each child has. The idea that first birthdays have more adults than kids in the photos isn't lost on me, and it's humanly normative. The intense hands-on parenting of infancy means that all-hands-on-deck generally turns into multigenerational adults extending presence around the young family. This is attachment at work in the same way that an infant turns into and cradles its mother when first cuddled. The orienting and attuning to each other is part of the adaptive attachment system that sustains our species.[7] We can do this individually as infants or adult parents, and we do this as a community when we gather to shower the mother or ceremoniously name the newborn. Bonding is a time when attachment is formed, named, and solidified.

BELIEVING

The work of the infant is to cultivate a developmental repertoire that prepares them for the relative independence, locomotion, cognitive maturity, and sociality that being a preschooler brings. As we shift from the terrible twos into ages three and four, cognitive, social, and emotional systems are different in function and meaning. Attachment specialists cite ages three and four as particularly meaningful in understanding later adult relationship patterns: The stories we tell ourselves about our ability and experience of love comes often from the stories we tell of our relationships at these times in our lives.[8] When our use of TSM is one that distances us, it lessens our access to understanding others well. Understanding others is a key task of Theory of Mind; when we don't develop this capacity, we are incredibly vulnerable to threat. A key advantage of Theory of Mind seems to be in predicting threat from others accurately.[9]

Likewise, preschoolers are also developing a sense of themselves and their worlds in other ways. Orienting toward others, demonstrating an understanding and anticipation of others' thoughts, and expanding on others' desires so as to meet a need are all actions that TSM facilitates. Theory of Mind is beginning to be cultivated in young children, and their engagement with and the risks and benefits of TSM for them is likely related to the degree to which Theory of Mind is able to be cultivated and nurtured in its formation. Evidence abounds, however, of the limitations of developing Theory of Mind in the context of technology adoption. Our ability to accurately judge and perceive others is impacted by our attachment. We are best at judging our social worlds when we are securely connected to others.

BEFRIENDING AND BELIEVING

If four-year-olds in their early childhood years must attend to the hard work of knowing what is true, not true, of them, not of them, and of the world, not of the world, we can forgive them for being perpetually in danger as they maneuver the broader world. Headlines of tragedies for young children often involve the age of a three- to five-year-old, and it's not simply because children are more ambulatory at those ages, but because the capacity of a child at that age dramatically changes, often before a family realizes the capacity has increased so much. If four-year-olds are the vulnerable wanderers of a given neighborhood, older school-age children are the "four-year-olds" of internet vulnerabilities. As reading, numeracy, and self-directed action abilities increase in children ages seven to eleven, the types of ways in which TSM is engaged change, and soon families find themselves worrying not only about the amount of screentime a young child absorbs, but also the types of settings on the screen that might limit who and how the child engages on screen. In my own household, and likely in many across the country, the TSM access conversations shifted from when to why, with some access being deemed too risky, too non-nutritional, or too mature for the child demanding access. How we parent around TSM for these ages changes too.

"Hey, make sure you don't give out your address when you're online!"

"Don't mention that you have siblings!"

"Don't respond to questions about what you look like."

"You use the computer in this room so we can keep an eye on things."

"Yes, I know your passwords, and I frequently log in."

Maybe you have said some or all of these statements to your child as they began going online more and more without you there. Maybe you've limited the websites your child can visit or the particular version of the platform that you allow them to access. And maybe you do frequently log in to check their various sites, scan their phone for new app downloads, and password-protect the family iTunes account so you know every app that's being downloaded. Maybe you've done all this and even more, or maybe you read that list and thought, *Wait, is THAT what you are supposed to do?*

Either way, I want to both reassure you (all good ideas, try a few, why don't you?) and point out that all that listed above aren't the most important things to do or say when you are allowing your child or teen to go online.

BUILDING

Now, to be honest, these skills are challenging to teach. As a parent, I'm somebody who wants to be the one who sets the rules, but what is also true is that when my children learn the skills, I will be increasingly challenged as a parent. The tradeoff to having a critically thinking child is having a critically thinking child monitoring your own parenting. Or if we are playing the same set of skills to others as service providers, then we will likely find they will be challenging our decisions more if we are being effective. This doesn't mean we have to change our values or the rules. We could be open to learning and understanding new perspectives, another scale we need to model, and why having a close and intimate relationship with our children is so essential. Parental leave laws and pediatric visits signal that critical developmental period of infancy, but adolescence is another critical period when the supervision that close family attachment provides has benefits for the risks and vulnerabilities of a given teen.[10] How much more time, care, and monitoring do adolescents need during what might otherwise seem like incredibly independent times, but that are, in fact, increasingly vulnerable times?

What we try to do is model how to critique and to always give our children the opportunity to fall gracefully back from a grave mistake. We need to actively teach forgiveness; we can do this by acknowledging when we ourselves make a mistake, particularly when that mistake concerns our child. Acknowledging our vulnerability, modeling humility, showing both vulnerability and determination to and with our children, can go a long way toward protecting them from potential manipulation. Have a talk with your child about how, no matter what they do online, you will help them find a way to deal with it. Do your kids know that you have their back? Does your child, whom you think is old enough to go online on their own, understand that if they do or say or talk with somebody there, that they can always tell you about it?

We do harm when we rely on media alone to transmit the cultural values we believe are most important. If we feel it's important to learn how to cooperate with others, if we believe that it's important to have ethical integrity, if we believe it's important to learn how to learn, if we believe it's important to recognize the value of second chances and mercy, if we believe it's important to embrace both creativity and persistence, if we believe people should be treated fairly, if we believe there are some things worth standing up for, then it's on us to have conversations with our children about these values. We need to teach them lessons about how we learned the importance of those values, and we must give them tools so they can actively critique the media. When it is giving messages that contradict those values, it isn't to suggest that we should be dogmatic in our approach, or that we should seek for our children to follow a perfect line of this, of a specific doctrine, but it is to say we should give our children tools to think critically and to examine closely the source and intent of media.

BRIDGING

What are the milestones that might mean something when it comes to navigating TSM use in our homes, schools, and society at large? Where does it need to fall in our parenting? In our policymaking, what might it matter if so much is out of the bag already? Isn't the message of babies that variation happens, and that we humans seem to move forward anyhow? All the babies, whether wiped with organic, fiber-free, compostable wipes or corncobs, seemed to thrive and giggle and take first steps and learn to coexist with those around them. Won't humans adapt and adopt their cultural practices in ways that meet their context, and so therefore move forward despite it?

As I delve further into the promises and pitfalls of technology adoption in media exposure, I see every day the effects, potential, and impacts of media exposure in children and families. What I don't see enough of are the conversations about how important it is to actively transmit your family value systems to your child. Or maybe what I see as comment wars take over online relationships during the political and social posturing between, I wish every parent understood how essential it is to actively teach your children the values and character traits that are important to you and that you use to guide your own behavior. If you are not doing this, you are making the choice to allow the lowest common denominator of online content to convey values for you—and ones you may ardently oppose. Media is a cultural transmitter; it introduces our children and ourselves to the norms, values, and standards of the culture from which it emanates. Media is a mirror, reflecting to us what we hold most dear. It is also being crafted to manipulate or explain a position or a behavior the general population should follow. Much of this media is ultimately about directing how we engage the economy. Media exists because capitalism needs capital, in some respects. So, because of this, media generally is the least common denominator of a culture and often is not excellent at conveying nuance.

Navigating TSM in adulthood and across one's adult life centers the ability of TSM to assist us in engaging the tasks we hold dear with adults; bridging culture and meaning across generations; and contributing to the community and individual legacy one will have in the world. TSM provides new options for ways this impact can be directed, and the technology touchpoint for us to consider is that across adulthood, the knowledge, judgment, and abilities we have begin in reverse order to be compromised. As vision weakens, our knowledge of what we see shifts; as that changes, our judgments about what we see also shift; and so, too, will our online behavior and interests. The vulnerabilities we navigate will differ as a result of our aging; and recent research is revealing the ways these shifts occur in surprisingly early mid-adulthood years.

Then what's left for us to do? If every stage where we are naturally oriented to go outside of ourselves or share what is inside of ourselves makes us vulnerable to harm and less attuned to our own impacts on others, how should we be considering TSM adoption? If such a tool is an unlocked door into the very central capacities that will make our children and our families strong, how could we ever think such a thing could be controllable, predictable, or amenable to guidance? What are we even asking to think it possible?

The question is a good one. In *Your Computer Is on Fire*, Mars Hicks and other contributors ask similar questions as they preface the collection of essays that lays out in excruciatingly well-researched details all the myriad ways our monopolized, weaponized, unregulated, and venture-supported TSM ecosystem is destroying our planet, democracy, and even the human biome.[11] What's the point of thoughtful reflection of privacy settings when state entities already have corrupted algorithms based on your loose settings in 2003 that will now dictate your social rating in a globally dominating TSM structure? Is there a best to hope for in such contexts? What strikes me about this is how quickly they escalate intimacy, competence, and intensity. The combination of those things, in an increasing dose, has an astounding ability to psychologically manipulate a vulnerable person. Even a person who might not otherwise be considered vulnerable could be created by these means. This is desensitization and deindividuation, otherwise known as Brainwash 101, and it is psychological abuse. It is incredibly effective, particularly for minds that are going through transitions in their lives, as an adolescent might.

I consider the next chapters simply a way to consider what is driving engagement and interest in TSM across life stages loosely aligned with culturally relevant ages in the United States in 2022. Maybe these insights will assist in developmentally appropriate expectations for online behavior and use. Maybe it will lead to focused congressional testimony on the mental health impact of influencers. Mostly, though, I think of what I am offering as *buffering*. In digital terms, buffering refers to the digital signal becoming consolidated and smoothed so that it transmits with minimal disruptions to sound and picture. It can also mean the storage smoothing that occurs in transferring a file. In terms of TSM adoption, let's consider technology touchpoints an attempt to buffer the negative effects of TSM, in our homes, our communities, and our globe. Technology touchpoints are developmental reflection points that we might use in crafting personal and public policy around technology and social media use.

· 7 ·

Bonding

$\mathcal{I}$ remember it fondly: first, the announcement that a new arrival would, indeed, arrive! Then the scheduling, the appointments, the extra people suddenly a part of my everyday life, even if it was only for a matter of a few months. THE WAITING!!!! The advice sessions, the urgency with which folks who already had celebrated their arrivals shared their best knowledge and tips to try to help me prepare for my own. It was a stressful time, and there never seemed to be enough information, although there was also a lot more information than I had bargained for. There was also the fun, the shopping! PINK! I could envision what it would be like to walk into that first big meeting after she arrived with that little cutie nestled safely in pink draped over my arms.

I am speaking, of course, of how I welcomed "Amy," a sleek, new white 2011 Apple MacBook issued to me after my tenure appeal defense was successful and my next round of campus tech had been approved and issued. I did, indeed, purchase Amy a cute pink woven business case to transport her most respectfully. As great as I thought she was, I met my then–department chair's consternation: Every inch I made to align the device existences in my life (I had by this time moved into my second iPhone, from which I had managed my hybrid work and family existences already for four years) raised his alarm bells about how to align department needs. His own coding and computer expertise gave him fast assurance that introducing non-PC users into his faculty team would derail the efforts he had made to align with hardware-driven technology innovations more common as the department technology was established. However, my own engagement with Boston-area e-media clinical researchers gave me the same, competing, assurance in the strategy I was adopting, and the college policy ultimately allowed such flexibility.

Fast-forward several years, and the original chair is the chair again, this time navigating even greater technology alignment challenges. I continue to use the tools, in their three years or younger versions, that I used a decade ago, and now I have my own team members challenging me on legacy technology and platforms that I ask us to move forward with new projects. Every new team member, with their domain and TSM-specific expertise and contributor needs, brings additional complexity. As a woman who has welcomed six children into my home, I know that anticipating any new arrival is often the same: the imagined arrival, the preparation, the actual arrival, and the network of people and places that become part of your life with that arrival, or with their departure. The same complex, intergenerational, multi-input reality is present when it comes to the bonds that are established through a family system adopting TSM imperceptibly via its engagement with searches, shares, and software updates.

The technology touchpoints linked with conception are many, interlocking, and network changing. Families, or individuals who are expecting to become a family, will engage with TSM in all sorts of attachment-fostering ways, both attachment to our human family system, and attachments to TSM systems and approaches that are forever challenged by the TSM attachments our family systems engage. As families and individuals prepare their journey to parenthood, whether through marriage, birth, adoption, surrogacy, or fostering, they will use communication to learn, and through stories they will come to understand, desire, and predict their future and their child's future self. Today's birthing circles include TSM just as much as they include nurses, siblings, and partners.

BEING AND BELIEVING

> When you understand that the pain of separation is healthy, normal, and inevitably a parental issue, you can learn to handle it... It's difficult, for your natural competition will surface at each separation and reunion.
>
> —Brazelton and Sparrow[1]

In the Brazelton and Sparrow *Touchpoints* series, the framework they outline explains the symbiotic back and forth a family navigates with their pregnancy and eventual child,[2] the prenatal and early weeks of postpartum are filled with the balance between the imagined future child and the actual pregnancy and infant. The weeks when a child settles into the care routine of a grandparent or sitter can be the same, as the above quote from *Touchpoints* attests. Brazelton acknowledged the emotional bonds that parents form with their child, even before a child is born, and after a child has arrived. As the unseen but felt becomes the seen and known, our relationship to ourselves as parents (the identity we develop over time of our own parenting self) and to our actual born child, shifts. This is due to the attachment features of the prenatal and postnatal environments, as well as the relational network the child and family forms.

A prenatal environment that is filled with stability and accessible resources, and free from violence, will foster idealized attachment, while one that is unstable, resource-scarce, or prone to violent or unpredictable stressors will foster an attachment environment of anxiety. These realities alter how the prenatal relationships and resources will be interacted with by the expecting family and its members. TSM will be part of that relationship by virtue of the resource it represents in accessing and establishing the environment into which the imagined child will arrive.

Once reached into, TSM takes on a network effect of its own, ushering the user and the associated accounts and digital footprints into tightly orchestrated and funded communication channels, themselves shaping the family system's idealized environment and child. The process is so organic it will usually be unnoticed. But eventually feeds formerly depicting video game ratings and fast fashion video streams begin to include maternity clothing, baby gear, and some public health information, iteratively shaped by TSM engagement across multiple platforms, browsers, and information portals. In seamless ways, networks of information about the TSM of the awaiting child's immediate social world are also brought into the common consciousness of the impending child's arrival, and, once arrived, become seamlessly brought into and cocreate the TSM of the child's early life.

The reason for this interwoven, informational-shared reality has to do, of course, with the specific hard- and software interconnections between device manufacturers, digital and social media platforms, and users. The reason, however, also has to do with the prenatal and postnatal attachment systems within the TSM; our current moment reveals that we are bonded through TSM in ways that are self-directed and representational of intended connection but also that we are increasingly bonded in ways that are orchestrated through moderation within TSM, either algorithmically or through direct human moderation. Our humanness seeks responsive relational engagement, even with TSM, and so we see the dance between baby and parent, between parent and TSM, and because of our humanness, soon we see it between baby and TSM.

PEEKABOO

The familiar chase-and-flee game that fills our memories of schoolyard games in our elementary years has its origins in the arms of our mothers and fathers. The parent, capturing the infant's gaze, and then releasing it, capturing it, then releasing it, and capturing it again to grins and gurgles, progresses by the time the child can sit up and motion with their hands into "peekaboo." A visual chase-and-pursue game eventually emerges between two or more people and sometimes pets. While Tronick's international work indicates that the pleasure of finding may not be universally experienced or expressed, the human interaction pattern and the playful modeling of what patterns are valued within a given cultural context is an incredibly important and human practice. When we see what we were trying to see, it's a great pleasure for us, and when we can interact with another using such a constrained behavioral reply, it's a great pleasure for the bond that forms between the players, as people.[3]

Images have a powerful effect on human emotion and cognition.[4] TSM makes images, along with their persistence, meaning, and transmission, possible. By the age of three, a child might encounter more than fifteen hundred images of themselves within the stores of their family's phone/photo books.[5] In the archives of our family images (digital and otherwise) is one captured during the summer just before the birth of my last child; my last fetal sonogram depicting the child we would welcome into our family. The tradition, one that began in my first 2005 pregnancy (lost within weeks of confirming our expected child), by 2011 had the ability to reach far more than just a few friends or family whom I might directly email. By the time of our final ultrasound, the sonogram was shared as a digital file, and I could use my new friend "Amy"—my cute white college-issued Apple laptop—to share it out with friends and family. The memento of our final pregnancy was a poignant one, and by then, uploading to Facebook was an easy addition to the friends and family email announcement. It's also a tradition with uniquely twenty-first-century roots and meanings, with some states even limiting the practice.[6]

I used the same laptop to put the finishing touches on my application for the Fulbright Faculty Scholar Award. Even though we were not yet the family of seven we would be when I was ultimately awarded the Fulbright, I had to indicate on the application form all the anticipated humans we intended to travel with if I received the fellowship. There, on the Fulbright application form, and then on social media and via our online submission, was information (sonogram photo, and a federal application listing their scheduled birthdate) about my child that existed via TSM even before I could carry the infant in my arms.

Images are powerful, indeed—of an anticipated baby, and of an anticipated place. As the Fulbright application process unfolded and we eventually learned that we would indeed need to relocate, the power of the image search came forward again. This time I was searching for other people's images of a Cypriot village (where my Fulbright would be posted), where a family like ours might have space and access to the universities where my research could be conducted. It was through these TSM-reliant strategies that we navigated the journey to the Mediterranean more than a year and a half after applying, and this time my anticipated child was no longer anticipated but actual. The village home in which we were lucky enough to live for our time abroad was offered by the clinical deputy within the ministry where my work would be conducted. I used Flickr, a photo-sharing site, and Google Earth, to track down the village and the old manor home we rented during our time abroad. It was a magical place and time for us, despite being away from our home base and despite the incredible cost for one of our employers, who navigated the internet access that we, as newly arrived immigrant visitors, would not have been successful navigating. Trying to establish a bond between our devices and our new country's internet helped us understand the intricacies of the public utility–based implications of access to TSM.

A LEGACY OF LOGGING

Leaving places and people requires an alignment of attachments. For humans, a place is a person as we generate identities about self and others within the contexts that we exist.[7] TSM enlarges the contexts within which human attachment exists, and as such itself will align as places and people engaging with it change. The rules and implications of the bonds we allow and enable between ourselves, family units, and devices within and due to TSM are often invisible. The last visible evidence many professionals will have of the ways in which they, their information, or their devices might be linked to TSM systems fades away with each smaller and smaller unit of technology and information storage we adopt.

By 2013, just after our return from Cyprus, and after knowing what early career work I would and, most definitely would not be returning to, I began working to disconnect myself from some of the earlier TSM adoption I had done, long before Apple introduced its iPod, but well after I was using Twitter both to conduct and to explore new research.[8] This liminal period, moving back from abroad and getting ready to move houses while also beginning to reconfigure my TSM alignments and approaches, is filled with memories of my own actual children becoming more of themselves, a process I documented, and as the TSM networks shaped by our devices, educators, employers, and networks extended that journey, it became documented in more and more fragmented ways.

Whereas in 2009 I ordered printed photos from my phone and placed them in books for my children to see for themselves, by 2013 I was moving files from various email accounts, online photo stores, and increasingly additional storage drives, both physically in our possession as well as exclusively online in cloud storage systems. Testimony about the various platforms documenting the journey to self has played out on forums such as Facebook, Snapchat, YouTube, and TikTok. Senate testimony from October 2021[9] reveals that while I was bonding with my newborn and using Facebook to search for potential photobook layouts, or when said newborn later shared their baby pictures in a TikTok video montage years later, little did I consider the ways in which I had bonded them, their user experiences, and identities, to not only TSM platforms today but to biometric databases ever more. When my husband and I planned our estate early in our parenting years, we had not considered having to handle a digital estate, and yet more and more of this became relevant for our children. When will the films and commercials depict the harrowing "remember, darling, when I die, you get control of the online family Facebook albums where I default-stored all your life images because my phone was always too full of videos from you playing with my phone" conversation?

Beyond the bonding of our biometric data to big tech is the reality that more and more invisibly bonded sources of information about our families and children exist and will be used. The future of medicine is the former future of business: big data. Through precision medicine, more individualized cures will be generated, and more of those will involve the combining of data with our biodata to generate holistic, and even networked, portraits of ourselves and our families. The reality of this future is actually present. Parents in China are no longer needing to navigate too much video game time for their children, as the country instead conducts this process itself, limiting access for those under eighteen.[10] Even users who might be that age will find their TSM use as one of many metrics bonding the user to some future access or lack of access, depending on how the social index will be used. Already as someone who uses life insurance, I understand how prior behavior is essential in predicting future behavior. But none of us are straight lines, and nothing is a perfect predictor—except, of course, the predictable ability of a few to affect the bonds of the many.

While devices bond parents to the access point they seek for reassurance, reflection, resetting, and registering, we now recognize the ways the TSM ecosystem they join, expand, and extend will have increasingly impactful reverberations as their children approach their independent adult lives, bonded to the TSM choices their parents or others made that now extends them far into the future. One day soon, instead of cancer cells taken from a living human being used to develop a host of human treatments, as was the case for Henrietta Lacks's uterine cancer cells and the claims now navigated by generations of her family,[11] we will likely hear of the first suit brought to claim earnings from interventions created through someone's media, medical, and biometric data. Who will make the claim, and what will this estrangement from control yield us in decades' time?

TECHNOFERENCE, ABSORPTION, AND DISTRACTED PARENTING

Bonding us and all future generations to our momentary relationship with TSM is not the point of early infant attachment, even if parents navigating that phase have brought TSM into their infant bonding and caregiving. The point of the human bonding that occurs in infancy is to promote the continued cognitive development of the infant through touch, eye gaze, dyadic conversation, and sensory integration.[12] As physical and cognitive development progresses, social and identity development also emerges. These imperceptible changes create infants who are prepared to communicate and interact physically in their social world. For babies whose parents use TSM—and according to Pew in a 2021 survey,[13] that is most parents in the United States—we can see their own pattern of TSM impacted by that of their family systems. As children develop their sense of self through caregiver interactions and sensitivity, it is through shared meaning and vision that much of the attachment work of these early years becomes established.

Three behaviors of caregivers to infants are of concern to researchers trying to understand how smartphones impact parenting: the amount of distracted parenting, TSM absorption, and "technoference" (having interactions interrupted by TSM use). Of particular interest in these studies are the specific ways in which parental smartphone absorption can alter caregiving behavior.[14] Increasingly there is a recognition of what I recognized in my own children as my use of TSM became more and more frequent. Juggling devices and being never not plugged in can mean a lot of changed caregiving practices since "screens in hand" certainly alter how we respond and interact with our "babes in arms." If the connection to the child is altered by the presence of a screen, how would we know? As remote working, new parents in 2021 and beyond will surely wonder, *Is there really much to do about it even if it is?*

The fact is, the studies investigating these connections are early, methodologically varied, and of a variety so that they offer no single conclusion. And even with the conclusion that can be offered (parental absorption alters the parent's responsiveness to their child), we know lots of things are also likely to do this, including chores needing to be completed, responding to work communication, or otherwise navigating the work-home balance typical in the modern world. What of the fact that any parental activity, particularly one for gaze coordination, face-to-face communication, and shared attention, is altered by these same issues: Does it mean that humans can only thrive with constant attunement? The key is consistent attunement, and TSM makes that harder for caregivers in many ways, particularly in ways that require touch, synchronicity, and response. That this is a consistent finding in the nascent literature is not entirely dismissive of the potential longer-term impact. Infancy is a finite period, after all, and children do eventually grow out of the babe-in-arms or baby-on-back phase, meaning that it's unlikely the earliest months of TSM use are permanently damaging to any given family system; however, increasingly those early attachments to TSM will predict the type of parenting attachment behaviors observed in each system or at a given time.

CROSSED WIRES

When parents of young children and infants talk about TSM, they are often not referring to their own use, but rather are fixated, sometimes via TSM, on their child's engagement with it.[15] Like most parents, we found ourselves not entirely agreeing with one another on the role or shape that TSM might play for ourselves or our children, particularly in the earliest years of our parenting journey. I had a firm sense that ours would be a home where children and parents viewed a common screen, the family TV, and in the early 2000s my progressive commitment was that our children would never have TVs in their bedrooms, a recommendation many studies were advocating when I was teaching to some of my first students in those same years. I hardly recognized the screen-in-every-palm reality that would transpire by the time we moved our children into a home with more rooms than screens. It was an easy set of steps to allow to unfold, in the name of fostering language skills and encouraging self-control. We and other parents are increasingly inundated with media content that is touted as educational and with messages that early engagement with media improves our child's school readiness and knowledge.[16] At the same time, we are admonished to monitor our child's technology access. It is this paradox that results in many families simply not actively navigating their family or individual stance on TSM. Instead, it seeps in slowly until its presence and the particular household realities of it are somehow formed and a child is joined into it.

During the early parenting years, when we scaffolded a single TV lifestyle while imperceptibly enabling a screen-in-every-palm reality, we could only rely on the Brazelton wisdom I could carry back to us via the training and certification process related to learning first the Touchpoints model, and later the NewBorn Observation Assessment. The theory and the test serve as useful interventions for any family system because they align infant development to the caregiving relationship, no matter the family system or developmental reality of the infant. Parents having their newborn assessed would be prepared for the types of soothing and regulating their infant would most respond to, and through the assessment conversation parents would experience the power of their human voice, touch, and gaze in parenting their new child. The NBO simply observes the child's behaviors and narrates to the parent what the behaviors are, how they communicate facts about their child's state, and how the parents can respond to support the child as they signal their distress or discomfort.

These alleviating observations empower parents to see their own observational capacities and enable them to create space for their infant to direct their parenting behavior, in a reciprocal dance known as "attunement." Preparing parents to know that their infants communicate behaviorally—and that parents can respond to that communication in positive, helpful ways—alleviates parent anxiety, making them less likely to allow their own "ghosts in the nursery" from driving too much of their parenting behavior. It's nearly the exact opposite approach I took when attempting to be supportive to new parents before I learned about Touchpoints.

In my early parenting, pre-iPhone years, it was more likely that we shared advice through emails. Reviewing my own from those early parenting years, I came across an email from a college friend, with other young mothers included on the message who once themselves were girls hanging out in dorms where we lived in college. As I remember these years, from the perspective now of someone writing advice on family TSM, I am struck by how obnoxious my approach was. In one reply I gave to my friend, who shared a book recommendation about sleep training, I shared without irony the advice I had, which was based on two children who slept easily and soundly throughout their childhoods, something sleep hygiene research affirms is not typical. To be fair, this note came after I had been a mom for several months longer than those on the "to-be" line. Still, I was insufferable, using TSM via email to provide arrogant, not amusing, suggestions that relied on a unique child and unique circumstance to implement, and that did not represent my own reality in any way, but did match the recommendations. In my reconsideration and reflections of these years, I can at least be transparent about the approach always taken in regard to my balance in adhering to my own awareness and options. Still, that's the thing about bonding with our children and ourselves as parents: We are constantly shaping our own and our circle's identity, and TSM can allow us to believe in our own wisdom, or to question our own wisdom and practices, in ways and rates that earlier human networks did not enable to the same, apparently permanent, degree.

In their second volume on the Touchpoints model, focused on children between three and six years old, Brazelton and Sparrow are ambivalent at best about the role and relationship of technology and young children. Briefly mentioned and situated as a technical modern family reality, their text offers both an endorsement on the role and importance of technology to support children whose capacities for verbal communication might be limited. Their mention of tech's child-enhancing opportunities is limited to its utility in increasing inclusion and accessibility of those children who might otherwise have little engagement options.[17]

> Children who have difficulties communicating or expressing themselves otherwise (perhaps developmentally disabled) can communicate via technology: "A disabled child with only his fingers to generate responses can establish his thoughts, his wishes, and his competence with a computer. Children who cannot speak clearly— or at all—can show the world who they are through a computer keyboard; and children who cannot use their hands can use voice-activated software.

Certainly, it's easy to see how such a general positive recognition translates into the recognition that remote learning is likely still learning, insofar as it provides an avenue for the attachment bond with peers and educators that contributes to children's learning. On one hand, they have some appreciation that it is present for families and even very young children, but on the other hand, their suggestions are profoundly underwhelming in their prescience or understanding of the phenomenological experience parenting with technology would be for families with young—but easily and perpetually internet-connected—preschoolers and children.

We might go so far as to consider statements offered on how families might consider the emotions that leaving a child at preschool—or with a sitter or some other caregiving arrangement, as one might say if they were reentering the workforce or adding support to their family system—impacts the system and the child.

But we all know it may matter to the baby in subtle ways, such as developing regular, reliable patterns of expectation. Adjusting to more than parental care is a challenge for the baby, though it may be an asset in the long run.[18]

In some respects, the observations and considerations could extend to notions of digital babysitting, and even in very young children like toddlers, parents will agree they see some emotional transition that occurs for children when shifting from engaging with a device to their caregivers' attention. Where Sparrow and Brazelton land with technology and children is rather underwhelming in clarity and in immediate relevance to our dynamic TSM era, yet it's a firm foundation that does provide some important lines of thinking that any family or system might consider in navigating the TSM of any era relative to the needs of human children, whose developmental trajectories and needs are rather static relative to TSM innovation. Within his books, Brazelton lists ten hazards and risks of allowing for technology in a child's life:[19]

1. Attention that demands physiological and motor passivity
2. The seduction of the computer as a babysitter when busy parents wish they had more time to be involved
3. Isolation from the rest of the world
4. Self-absorption, and the lack of rewards shared with others
5. Interference with learning to communicate face-to-face and to understand peer relations; time away from peers replaced by dreaming and fantasy in a virtual world
6. Distraction from or even loss of involvement with other activities important to well-being, health, and development (Social and physical activities may be missed out on, or avoided if these are challenging to a particular child.)
7. Overeating (as with watching television) if food is permitted at the keyboard
8. Poor posture, eyestrain, and strain on hands, arms, neck, and back—all these are risks of prolonged repeated use (They are aggravated if seating and positioning of equipment are not ergonomically appropriate.)
9. Possible health hazard of electromagnetic radiation from television and computer monitors (These remain unconfirmed and controversial.)

10. Immediate rewards (The computer's capacity to gratify a child
    unfailingly can be a drawback as much as a boon. Children
    also need a chance to be frustrated and to tolerate frustration,
    to delay gratification, and to find satisfaction within themselves
    for their efforts.)

Ultimately, we must navigate together what children's technology and development pioneer Chip Donohue advises when we speak a second school year into the COVID-19 pandemic: "How are we using media in our home? What does it look like here? What's digital well-being mean for us? What does living well with media mean for us?"[20] Families will likely define that in ways unique to their own attachment patterns and attunement.

While the infant and family system might universally benefit from anticipatory guidance and an emphasis on relational attunement with their child, this current generation of parents of young children is likely the first to find themselves growing up with an online presence since early childhood, and within a few years from now, the first online-always generation of parents will begin raising their children. The ghosts in the digital nursery will likely be many, but for now, while we wait more long-term research, and while we assess and observe current families and children navigating the bonds of TSM, we do well to rely on the basic premise that human development has function and trajectory, and that relational attunement, early and often, is the critical foundation that enables the physical, social, cognitive, emotional, and sensory development that humans require to become prepared to navigate a world for their own survival and reproductive ends.

SOLIDARITY AND SORTING: TSM'S BOND

By the age of only a few months, babies can form opinions and predictions, and they don't hesitate to make it known when their predictions are not confirmed—although a parent staring at a screen might well miss the obvious, if subtle reaction. This ability to form understanding and opinions so early comes from our human need to secure our spot within a social network. Babies come primed to assess the intention of someone interacting with them or others in the social environment. Even seeing a puppet give an extra grape to another puppet can trigger opinions of likeability and fairness in infants as young as eight months old.

This observation and prediction ability stems in part from an emerging Theory of Mind (TOM) that the infant acquires through development, caregiving, and experience. Through this emerging TOM, children over several years develop the capacity first to appreciate that there is great variation between what I might want and others: This Diversity of Desire understanding is the first aspect to emerge in a TOM for children. In some children, the next element to develop is an appreciation that different people KNOW different things. The Diversity of Knowledge aspect of Theory of Mind emerges as children increasingly navigate physically without their caregivers, as it is the child's emerging awareness of their knowledge that increases their ability to then use this awareness to meet their desired aims.

Suddenly, then, as the infant becomes the toddler and as the toddler becomes a preschool-age child, they will develop False Beliefs, and this aspect of their TOM will enable them to create fictions, fictions that can be used to obtain a variety of attachment needs. For it is only as children develop an understanding of false belief that they can lie, pretend, or create stories, and without the ability to do these things—children's capacity to believe in themselves or to identify true from false—is absent. TSM may not derail TOM, but it does provide important context about the child and their family, and many families capture and release some of their children's earliest TOM display, like sharing adorable videos of kids using their toys and household objects as pretend smartphones and mimicking their parent's unique TSM interactions.

In Chapter 8, we will consider what this means for children's emerging sense of self and the ways in which these influences converge in early childhood as families move from bonding with their actual child to helping that child develop a set of beliefs with which to navigate their increasingly social and imaginary worlds.

· 8 ·

Befriending and Believing

"*Close your eyes . . . am I still with you?*"[1]

The woman sharing a story for the national StoryCorps archive wakes me, her voice in my ear from the Alexa speaker near my bed that alerts us with our local NPR station each morning. She is quoting how her mother comforted her as she faced her own death and her daughter's grief over losing her. The poignancy of this, how the story illustrates the way love makes our attachment objects permanent, starts a tear in my eye. The same kind of object permanence that a mother leaving her child hopes is possible no matter the age, or temporality, of the departure, is a moving reminder of our minds' power to make meaning and to anticipate reconnection even when we can't see things with our own eyes. That attachment relationship and the Theory of Mind cultivated from infancy becomes the child's foundation for its most essential developmental work: pretending, befriending, and believing, which are the developmental tasks and activities of early childhood through early elementary school.

Object permanence, social cognition, language, story, and meaning bring connection, and these bring coherence.[2] A child between seven and thirty-seven months traverses an extraordinary gain in their sense of self through the development of these abilities, as Theory of Mind expands and deepens for a child and as a child's relational network forms multiple mental models requiring fluency and benefiting from rehearsal and engagement. In these months, TSM use, which a child's physical and cognitive development across their second through fourth years allows them to expand—for both the child and their caregivers—can support connection. In the case of families stressed and with disconnection within the family system, it can interfere with relational attunement supports that can assist in meaning-making at a developmental point when a child's verbal skills and self-regulation skills are primed to emerge.

Experiencing a lack of relational attunement due to hyper-engagement or digital babysitting can create disconnected children, slower to develop self-regulation and emotional attunement, although a household with this pattern of use likely already has other challenges going on that present a risk for the child: Assigning the risk of TSM interfering with a child's development is therefore a tricky thing to assess in a lab. A desire for TSM seems universally attractive for parents and children in these stages. The implications of their being focused on a digital device might be understood and guided with attention to the system and the attachments functioning within the home. When TSM can be used to support a child's games of pretend and imagination, then it can have benefits for children in deepening their reasoning and visual processing skills. At the same time, extensive engagement and excluding engagement (where other play and communication behaviors are traded for TSM) can be detrimental to the relational and self-regulation skills that are otherwise prepared to emerge.

While there is no single linguistic sensitive period, there is a sensitization and social cognitive benefit to linguistic proficiency accruing between two and four years; the infant's self-regulatory capacity is enhanced with the framing, meaning, rehearsal, narration, and regulation that language provides a child.[3]

> Children between two and four not only learn to talk but also learn the ability to communicate ideas and feelings. Recognizing a child's inability to speak before the age of three can be critical to her future social development. At this age, a child who can't speak is also missing out on language as a way to make her needs known, and to discover that she can have an effect on the world around her.[4]

Redirecting parent TSM engagement, depending on the degree to which it functions as avoidant or anxious attachment sources, can support parents and help them see that they, too, are personally navigating TSM in increased ways often as a way of supporting their connection with their children. TSM can be a soothing or stimulating support for a family, but outside of the caregiver lap, it may be diminishing a young child's emerging ability to meaningfully process information and even their developing self-belief. Not enough interacting time with a young child can deteriorate their capacity to relate to others over time. As children begin to communicate verbally, they have begun to attach differently. Children progress from relying on parental attunement to facilitate state transitions (between sleep and wakefulness, hunger and fullness, calm and irritation) to relying on their own ability to influence their state and their transitions (by communicating verbally, ambulating, and the use of social cognition). This is related to the impact that language acquisition has for emotion processing and control. For children who have speech delays or developmental disorders, such as autism, there are interventions that are designed to support their development of relationship-building behaviors. Applied Behavioral Analysis is a type of behavior-shaping treatment that rewards children with relational or language delays, as seen in autism, for positive visual and social responses to others. When provided during late infancy to the preschool years, it can dramatically improve the ability of a child with communication difficulties to express themselves verbally and to form social connections with peers. There is a profound organization of their relationships and social world in the child's mind that is established because of their verbal capacities as they emerge in these years.[5] While children who cannot speak will still communicate and develop emotion regulation with supports, it is a strength and a protective factor to have verbal communication skills and cultural contexts that affirm and attune to them from these preschool ages and forward.

By the age of four, children's concepts of their attachment figures begin to deepen as well, and in adult attachment style theory and interpersonal therapy techniques, these preschool relational patterns are predictive of adult relational patterns for many.[6] Those with connections that are secure, stable, and predictable (which around 60 percent of people in infancy and toddlerhood experience) will have secure and relatively stable adult relationships, and those with anxious attachment (uncertain emotional response from a caregiver, pursuing and distancing behavior in the child) translate into anxious, insecure relationships in adulthood, characterized by feelings of jealousy, isolation, or self-blame (about a quarter of adults) or characterized by avoidance of commitment in the adult years, tracing back to avoidant or disorganized attachment in earlier toddlerhood.

These are not perfect correlations, of course; both environmental conditions and individual experiences can alter the child's later adult relational patterns.[7] Still, the patterns of attachment we have as we parent children in this age can give us windows into our own prior patterns, can highlight the negatives of a prior relational pattern, and can highlight the opportunity at this age, and that the threat or promise of TSM may hold for it within a given family, or for a particular technology. The preschool developmental window is all about the child incorporating abilities previously reliant on the external and to gradually offload the cognitive effort required to master the various capacities that quickly erupt across these months.[8]

TSM can act in a similar manner for a family with young children, serving as a capital good[9] in the way it makes possible capturing sweet moments, gathering important information, and connecting with others, enabling a family to offload effort that can enable greater creativity and lasting memories (however misunderstood or misapplied this might be in capitalist contexts). We need to recognize the pattern between children, their caregivers, and the TSM they adopt.

The core of connection is understanding. Story builds understanding by creating shared mental models of objects and concepts we otherwise have difficulty holding fixed. Narrative allows our minds to do what they most want to do, which is to offload effort. A child being able to learn a story is a child who is able to connect to others, by sharing, processing new information by extending from prior stories, and by recalling connections to important others through story. If genes are the code that writes our futures and pasts, stories are the code that shows the past and shapes the future. And so, this is where TSM will enter largely for families. As the glow of the mobile device so fascinating to sensory-driven and drooling toddlers is the parent's smartphone, it's the family tablet that has our preschoolers' attention. Appreciating the way in which tablets as a tool are a capital good in homes with preschoolers can help us understand this rush to reassess where we must attune expectations to design features, and features of the child and their family setting. As we do this, we might better appreciate who is capturing their attention, and the way in which it is shaping their attachments.

CAPTURED AUDIENCES

YouTube's massive collection of content contains a variety of videos that easily engage children, as many feature children or feature items preferred by children. The company, now owned by Alphabet (Google's panoply umbrella of tech businesses), seems to be attending to its own "stickiness" and the impact this has on youth; it set up "YouTube Kids" to provide more limited content exposure to younger audiences, allowing parents to further refine their criteria through setting controls. However, despite lots of feedback and disturbing examples of inappropriate content before the 2015 introduction of YouTube Kids, the primary YouTube site provided few automatic protections. While the protections that a parent can select may be helpful in limiting exposure to graphic material or to enabling or disabling the auto-play feature that cycles from selected content to algorithmically derived and pushed content, and while progress has been made to allow parents to rest assured that children's content views do not collect the same degree of user metadata that happens to their data when they view a video on the main YouTube platform that is not children's content "gated," none seem suited to controlling the product exposure faced by young children engaging with videos.[10]

The platform does not allow ads on the YouTube Kids platform, but when the content in the videos themselves are influencer-placed ads or children attempting to yield enough of a following to become formally paid for advertising specific products, it hardly matters if there is a formal ad program. The children's videos that my children found easily readily focused on toys and candy, then gymnastics, dancing, trick shots, gaming, slime, anime, and business shops of all kinds. Inevitably the content online was easily demanded within the shops we frequented. Preschoolers soak up information and particularly like to mimic adult motions and cadence, so the songs, dance videos, and musical content that entertain a toddler necessarily move into short stories, and in particular stories where the hero adventures. For a preschool child, connecting with a hero, their hero or special friend, can become a lifelong memory. Imaginary friends or mini-fandoms of favorite characters in a story, web series, or movie are common. Our love—for, say, Elmo or a K-pop band—is an example of the parasocial relationships that begin to emerge in these stages.

My visit to Sandra Calvert's lab described in Chapter 2 was focused on TV-mediated relationships, but more and more I observed children forming these parasocial relationships with real children, who presented themselves as characters in the YouTube channel landscape my children consumed on our family TSM. It made me wonder if the families of those children understood the parasocial bonds their family was committing to as their videos gained viewers, or if the same eighteen-month timeline for intense engagement translated fully for parents and their YouTube popular youth. The scale and spread of the channels and their enduring availability enables these qualities in unique ways compared with the produced televised content that most children's media safety laws are focused upon.

It would not be long before even that digital rehearsal of their favorite video makers became the point of my children's pretend play. By the time we meet Ben and his grandparents at the beach years later, my children's filming of their playground experience was their focus, while the four-year-old girl I spotted in the corner of their video simply wanted them to see her—the need to believe pushing her to engage her near-peers while they sought to engage their imagined peers and future re-viewing selves. The other rehearsal they would engage out of these preschool years was that of the consumable items the content creators they tuned in to preferred. Conveniently, these products, or products to create the slime in neon colors, and to store the slime, and to promote the slime conventions that would then pop up regionally a few years later, pointed the early converts to slime videos, and now rehearsed the entrepreneurial channels being fed into their feeds now that their age demographic had progressed. To support this moment, deliberate choices can be nurtured to shape and support the content taste preschoolers are establishing, but the most essential aspect of the TSM engagement will always be the narration opportunity it provides the child, and the opportunity through symbolic play that the child can engage, either without the device, or through it interacting with those social and parasocial relationship targets it loves.

Any parent who attempts to hurry a child along in ending their parasocial attachment will certainly convey how challenging it is to set limits with children concerning anything to do with their favorite character or show.

Media-mentoring expert Chip Donohue[11] is pensive as he considers the battles families wage in measuring the right dose of early tech engagement. When we speak in the fall of 2021, he is still operating in the remote working world of the pandemic from his home in Wisconsin. It's a week heavy with headlines of hurt communities considering the fates of authorities and self-appointed authorities across the country. He seems hesitant, like so many, in weighing in on judging parents for their best efforts.

> Those of us who are very privileged can talk about controlling screen time, and managing, and giving other alternatives. And maybe that's not an option, and maybe during the pandemic in particular, some families have faced a situation where they, they have to . . . The screen is the way they're going to survive that hour, or that moment, or working from home, or whatever.[12]

He is right, of course; this is what families navigated during the great juggling act the pandemic raised. His approach that parents would ultimately do their best was different from the ambivalent admonishments Brazelton and Sparrow raised in their work, writing that

> Parents also need to be familiar with the content of software and internet sites a child is using. They will want to shield their children from violent computer games. They are unlikely to help children work out angry feelings, nor do they serve any other constructive purpose. Violent images have been shown to induce aggressive behavior in young children, and although some children may be able to manage this sort of overstimulation, others may become overwhelmed and preoccupied. Some are so frightened that they try to compensate by imagining themselves capable of violent acts.[13]

What seems out of place for today's parent perhaps was a bit more obvious to the parents he was originally communicating with. Parents in the early 2000s would have navigated a barrage of dress-up toys for their preschooler, while parents just twenty years later are more likely to confront ads for digital apps or nature-based abstract toys focused on sensory aesthetics.

POWER OF PRETEND

Venturing out into the wider world is the tricky risk of a preschooler, one they can brave armed with the transitional object of their family and group: a car, blankie, a stuffed animal, a tablet. Whatever the familiar item is that can provide the cue for soothing that the child seeks when distressed will become important to the child, and because of its power, to the family as well. While those types of risks and protections were familiar concerns for parents with young children, the concerns I began having with my own children's tablet engagement in their toddler years had more to do not with content exposure, but with the lack of story-telling their story consumption was generating in their own play behaviors. Within my home, tablet and gaming use led to *less* venturing when it came to physically traversing the neighborhood, or playing in the woods near our home, or in using words with the grown-ups nearby.

Dressing up, pretend restaurant, puppet shows, rugs with lines of cars ready to race, dolls ready to be cuddled with care—this is how my children born between 2006 and 2009 played in their toddler and pre-school years, the years when social cognition and Theory of Mind were emerging and consolidating. They physically manipulated toys, re-created rooms with their blocks, and engaged in the physical but symbolic play that is characteristic of these stages. Their toddler years were before iPads and YouTube content was easily accessible.

In contrast, my children born between 2010 and 2012—and my nieces and nephews born even more recently—engaged in digital play, scrolling content videos, favoriting repeats of *Odd Squad* on the PBS Kids app, and then using a digital phone (one active like my smartphone, or one of the now-abandoned but still-functional older iPhone models we had by then added to the device inventory of our home). Their rehearsal, occurring in the 2013–2016 years and beyond, was far less physical, even with older playmates available and accessible in the home.

Instead, their solo and social play was often a digital rehearsal, a storytelling and meaning-making that incorporated a social cognition of "viewing" actions, and of that re-viewing. Taking advantage of moments when videos might replay was the key to capturing a moment to hear them narrate and share their work or their interest in someone else's video. Where my older children might have used the camera feature of whichever parent's iPhone they could smuggle away, my younger children used it to capture their toys and narrated scenes, often mimicking the scripts of the YouTube child they most often viewed.

Preschoolers are a swirl of recognition and rehearsal. The world is a place where things can be done, so they will and should be trying to do them! TSM engagement can be so all-consuming that the physical and role exploration that comes from active and social play will be compromised by sedentary and passive consumption when brains and bodies are developmentally oriented toward doing.

All that doing is likely to lead parents to lots of capturing and sharing. By the time my own children became preschoolers in 2010, a new term had already emerged: *sharenting*.[14] What Brazelton would encourage as normal, universal feelings suddenly were time capsules of parent musings and positions on their child. It would not be long before those children would protest the practice, even as their preschool selves became enamored of their selfies and videos.

MAN ON THE MOON, MUPPET ON THE STREET

It wasn't really the novelty of my preschool and kindergarten children reenacting in video form the videos they themselves consumed that was striking enough that as I observed it, I challenged it. Rather, what I observed was something that had already been well-established in the psychological literature. I wasn't shocked to see my children's play consist of playing as though they were themselves YouTubers, and I wasn't shocked at what channels on the platform they would "like" and to which they would "subscribe." Albert Bandura's classic 1961 study (involving a nice, young woman, a hammer, and an inflatable clown named Bobo)[15] had already prepared me for the reality that young children are keen and astute observers and re-creators of near peer and adult behavior, and they are highly motivated to replicate it, especially when the one they observe is showing excitement, intensity, or otherwise animated and engaged themselves. These are the exact prescriptions for a video with many views if there ever was one, and, in fact, there were several in the forms of algorithms that would yield increasingly engaged, and sponsored, content creators, directly into our YouTube feeds.

The social zeitgeist that was part of this era included the racial and social inequity tensions our own era continues to struggle through. In these early years of children's television, Senate and Congressional hearings were held bemoaning the "vast wasteland" of television programming for *any* age group, as well as a concern that mass media might not be amenable to the prosocial needs of young children.

In part, what I was noticing (as my children became pretend YouTubers) was the same process that captivated Bandura and was part of the research that emerged to understand, and eventually counter, television media as it was becoming more and more of an influence on what children might do, think, or believe. In the 1960s, there was both elite aversion to television adoption and use with children, and a great series of revolutions in the combined power of creatives, child developmental scientists, and foundations seeking to improve the quality and impact of televised media, in particular for children, and in explicitly public-funded, non-commercial ways. In the 2021 book *Sunny Days: The Children's Television Revolution That Changed America*, author David Kamp[16] outlines the mix of Kennedy New Frontier idealism and the Johnson administration's legislative support for school integration and reducing rural and urban poverty that resulted in the creative contributions of Fred Rogers, who created *Mister Rogers' Neighborhood* in Pittsburgh from 1964 to 1967; Jean Gatz Cooney, who would create *Sesame Street* and form the Children's Television Workshop between 1967 and 1969; and Jim Henson, who would take his Madison Avenue sensibilities and his own maker-ethos and create a world of street scenes and variety shows for *Sesame Street* and eventually *The Muppet Show*.

The children's television revolution might have ridden the technology innovation of the Kennedy Space era, and the Eisenhower post–WWII broadcasting and social science infrastructures,[17] but they were being led by serious, public-minded creatives who saw the profound power of the medium and who did so at a time when a revaluing of young children's great social and emotional capacity was being recognized and nurtured. Rogers and Cooney demonstrated that with greater attention to the needs of young children, and the cultural context of their lives, media was able to improve health outcomes in their culturally diverse and impoverished families to a meaningful degree.

Families facing the challenges of poverty, mental illness, addiction, and cultural adjustment have all been identified as also experiencing challenges in effective TSM access, ease of communication, and TSM limit-setting with their children. These difficulties often follow children into their school settings; children of families with these disparities more often experience truancy, behavior problems, and lower academic achievement. Risk research offers only one part of the equation; protective factors that stem from family cohesion and school connectedness can be fostered and cultivated and in themselves offer increased resilience for families at high risk. TSM can play a unique role in supporting these families, through content that supports and anticipates the parenting and limit-setting modeling or content opportunities that can support preschool parenting and preschoolers, or in connecting families to their local networks, enabling the development of a cultural setting for their children's most social period: elementary school.

· 9 ·

Becoming

The Magic of Middle Childhood

The early power and prowess of preschoolers' make-believe yields a belief in themselves, as dear Mister Rogers so wonderfully demonstrated in his televised programs for young children and in his support for those who teach them. Belief in yourself, he understood, yields important exploration, enthusiasm, and engagement that lack of self-belief would prevent. The early years give way to the middle years of childhood as their Theory of Mind develops; their capacity to engage not only in imagined play, but in actual work, increases. The contribution to family and groups that children make in these years can greatly enhance their family life. It is this capacity, and the distortion of it, that heightens parental and policymaker concerns about TSM.

The recognition that early viewing of television might alter behavior in children spawned a host of media-focused research throughout the 1960s until today. The studies of observational learning that brought the phrase "Bobo the Clown" into General Psychology textbooks evermore was one of many in the decade of the 1960s that focused media and policymakers onto the impact of television for children. The burning questions of whether being on film year after year or viewing the results of film hour after hour in a televised broadcast might impact a child became of great concern by the end of the 1960s, when social unrest revealed the continued effects of "separate but equal," and urban disinvestment along with housing redlining contributed to poor educational opportunities. The critiques and concerns of television in the 1960s concerned themselves then with the program impacts for young children, and the efforts they made created an entire new industry of edutainment.[1] Within a few years of their efforts to promote programming for children, the nation's first report on the effects of televised media on children was released. Its conclusion? Some media, for some children, can cause some problems.[2] But which, when, and why? That thinking would play out for the next five decades.

Even federal deputies in charge of regulating airwaves in the early 1960s bemoaned the impact of television on Americans, and, in particular, children. A thoughtful Fred Rogers, however, felt there was a special and important possibility for television, particularly for children. In that decade, a minister, a social scientist, and a puppet maker empowered by Pennsylvania and New York media affiliates were forces that would change the way televised media was thought of and used with young children. Fred Rogers in Pennsylvania, Jean Gantz Cooney in Manhattan, and Jim Henson in Brooklyn were pioneers in promoting the potential of engaged, contemporary, and multigenerational television, particularly for urban children who might otherwise lack the opportunities for learning that more rural children would encounter.

By age five, preschoolers enter a period in which they blossom on their way to adolescence. They develop greater capacity for spending time away from the family environment and become increasingly connected to external environments, such as school, faith communities, and the broader community context. This stage, referred to as middle childhood, juvenile phase, or (for the later ends of this stage) pre-teen, is the age that most of our beloved children's literary, television, and cinema characters depict; children in this age band are beguiling in their capacity to be children, while displaying sophisticated and culturally inquisitive observations and actions.[3] These are the ages and stages that adults often look to for examples of precocious support, raising funds for veterans, or across human history raising siblings through mutual care and support as members of family systems. It is the stage of life that involves two essential tasks: being developed with cultural knowledge and being integrated into cultural spaces. These tasks of social learning and integration are the caregiving and developmental "work" of the middle childhood years. TSM, like television before it, and like radio before that, has a particular impact on the years when children first become both receivers, and, depending on context, providers of support within their family systems.

What makes this possible was referred to by Jean (and Valentine) Piaget as the 5–7 shift: the developmental years when cognitive capacities increase and enable the child to achieve a greater degree of abstract reasoning, while also unfolding during advancements in their physical dexterity and capacity to learn terrain, topography, and distance in ways that enable exploration.[4] Physical dexterity, cognitive flexibility, and a developmental sociability all connect in the years as children move from learning to engage the reading and calculation skills of their environment to employing them in the service of their own socially and personally relevant goals. And those goals become increasingly social goals, ones focused on attachment; however, this attachment is now focused not on the secure base, but away from it. Parents and important adults generally move from the center to the periphery of novelty and importance. Literacy and numeracy skills are dramatically increased and solidified during these years, with most research now recognizing the cognitive shift in printed media that children navigate; moving from learning to read in the early childhood years to reading to learn by around grade three and age eight.[5]

Why would this capacity develop at these ages? Evolutionary developmentalists opine that gathering food, caring for younger siblings, and otherwise contributing to the small hand work and sustenance farming tasks associated with family life are advantageous to family systems, enabling adult reproductive advantages and bolstering the value of the child to the community of which they are a part. Important brain regions associated with disgust, and that might provide protections from foodborne illness, are not yet developed.[6] This may make engaging in some food preparation tasks possible with less distaste and enhance the success of the family system (of course, that lack of disgust is also a risk, and as we will explore in Chapter 10, it's a pivotal factor in moving the child closer to their adult status).

While television was busy becoming, so were the Boomers. The generation of children born and raised after World War II is the first generation for whom televised media became part of their everyday lives and interactions, and the explosion in music, youth culture, and fashion have forever been altered because of its ascendence. In the United Kingdom, Michael Apted followed a group of young Baby Boomer children over their lifetimes. One of the most intriguing portrayals of the "now and then" query comes in the form of a film produced for the BBC directed by Apted, who in 1964 was a newly minted film school graduate working for Granada Television in Manchester, England. At the time, a senior member of the television staff had an idea to film schoolchildren and to examine the social circumstances in which each child found themselves. Tim Hegrew, who produced the documentary, was fascinated by the profound class differences in London's children's experiences and sought to document what kinds of life opportunities would appeal and present for children raised in widely different environments: from rural villages in the Scottish Highlands, to council estates in East London, to the rose-covered cushions of an elite private boarding school. The documentary grad, Apted, signed up to help with the project and found himself working with only three weeks to cast twenty-five children for the production. Of a motley hundred or so children, he crammed interview drives into a tight three weeks; from Sheffield to Dover, about a dozen families allowed him to capture their child at age seven, and then every seven years thereafter.[7]

At the time, in 1964, the idea that Apted worked from, that something of a child at seven is permanent and fixed, seemed a given. Nature, after all, was what determined character, and character certainly determined one's life. In an interview with Apted a year before his death—and following a screening of *63 Up* at the Lincoln Center New York Film Festival—he revealed how he might have championed television, but for him peeking into the fixedness of a given seven-year-old was not really his endeavor; rather, he understood his endeavor to be about revealing the harshness of Britain's class fixedness. The technology of film and the ability to remain in contact with his subjects created the perfect conditions for Apted, and us, to reflect on the impacts of exposure on individual personality. What fascinated the UK television audience, and what excited Apted about his first job in television, was the engagement of seeing something—someone—"BECOME."[8] In 1960s England, televised media was, as it was in North America, nascent. Most homes did not have a color television set, but that would change by the end of the decade. The program *Seven Up* became a cultural and then a global phenomenon, and every seven years the filmmaker and a crew from BBC would interview as many of the original fourteen (ten male and four female) children who were willing to continue their participation: Suzy, Bruce, Neil, Paul, Jackie, Tony, Paul, Simon, and the rest who were still participating decades later.[9] Becoming requires human connection: It is through others' social views that we determine our own worth and potential. The very act of appearing on television at a young age might be expected to shape a child; indeed, when I encountered Apted and a group of now-adults who appeared in *35 Up* as a student in Dan Mrozek's personality psychology class in the late 1990s, we wondered about the implications of growing and becoming before a camera's eye, particularly in a small community with large global impact, which a television broadcast from the United Kingdom would be throughout the Commonwealth. In actuality, and in consequence, TSM, like television before it, enables even further distortion of the identity development process.

Eventually the seeing and becoming requires bonding to identities and groups, since being human requires other humans in some form, even when remote or distant. Just like the parents of Suzy, Bruce, Neil, and Nicholas couldn't have known what they were opening themselves and their families to for five decades, a single app download or group hangout online can have similar lasting impacts. The very questions Apted's documentary explored are now, finally, regularly emphasized in psychological science studies of development and recovery. Seeing, believing, and becoming require representation, valuation, and identification, processes altered by TSM through algorithms and user behaviors. While the ambitious project certainly has impacted Michael, the British public, and its own sensibilities about its class structures, it is perhaps the impact the filming had on the children themselves that viewers most appreciate between the films.

What of the impacts on the children themselves? The modern viewer can ask these questions and get contemporaneous reflection from Neil, Bruce, and Tony themselves. What did inadvertently becoming a cultural icon for life mean for the individuals who, within three weeks, had unsuspectingly opened their lives to world scrutiny? These questions are at the heart of this book to a large extent: What impact will opening the world and our children's lives to its scrutiny and cloud-stored memories hold for our children? What would keeping each from each other mean for them? Mrozek's work, then at Fordham and now at Purdue, focused on change and continuity in personality, and we engaged in many articles and research in the course. The highlight, though, was viewing *35 Up* together and then using the earlier productions to see the continuity and change research on display. In the course, and in my own life, we engaged the opening question of whether a child at seven could show you the man at seventy. What was it about age seven, or age nine, or age eleven, that such preferences are still salient so many years later? I am not unique in these regards; the timing of these years coincides with an evolutionarily advantageous exchange of culture and labor.

This vital stage of childhood also serves a vital cultural function, and changes in the historical attention, emphasis, and elevation of this age group signal a socially relevant desire for optimism, buoyancy, and connection. If there were a sensitive period for social connection and positive community attitudes, this is an age that rises to the top in importance. This might be positive and playful, but at a social level, this likely looks agentic, aggressive, and assertive in its energy and positioning. Later childhood is about creating one's world and deepening one's abilities. These qualities about the juvenile period of human development, which precedes the adolescence phase, are critical to their attachment and affiliation to a group. The magic of middle childhood rests in the cultural transfer and transformation children this age embody. Sociality dramatically increases during these years; peer engagement is sought with urgency; and tactics like forming relational alliances, displaying talents of value to the group, and engaging relational or physical aggression, for individual or group ends, increase.

This combination of interests, proclivities, and tactics has a purpose; social competition serves central needs for this developmental stage in areas of social cognition and self-regulation. Whom we view as competitors results in our sharpening the skills we associate with them, and this collectively, within a community or group, can yield improved performance for these skills as a result of the recognition, replication, and reward cycles held over from our earlier observational learning skills described in Chapter 8. Ask any eight-year-old if something is funny, and there's a high likelihood you will learn who that child's verbal sparring partner is at home: Humor will develop in relation to social references and in response to the social and cultural relations of their caregivers.

As the cultural transfer occurs, children are also responding to brain developmental paths encouraged by the adrenarche period, when adrenal glands begin to contribute to pubertal precursors. Children in this stage tend to gravitate toward same-gendered playmates and an interest in adult gendered roles, evident in their play and communication. During these periods of development, earlier attachment patterns with caregivers activate attachment behaviors, with different early attachments corresponding to middle childhood attachment styles; boys moving from insecurely attached to avoidant in their behaviors, while girls with insecure early attachment moving toward preoccupied or ambivalent attachment patterns. [10] Rather than representing patterns of actual interaction alone, these changes are also related to prenatal hormone exposure, something highly subjective to stressors experienced prenatally by the mother.

In family, peer, and community circles, the child is attuned to the whole and is learning and practicing the norms, roles, and opportunities of the community. This exploration is beneficial for family systems, increasing their network through forced connectivity informally, and formal connectivity from formal interactions associated with juvenile transitions into adolescence and social roles. However, when this goes poorly, and children are harmed, we often see TSM remediation that emphasizes children's safety features, even when this is precisely the time when the technological capacity of the child outpaces their parent in terms of features, settings, and updates. In this exploration, tech has a specific role and risk, and guidance around adoption has as much to do with individual child needs as it does with the limitations of a given system, family, or network.

For all the Congressional bluster on sanctioning food advertisers back in the 1990s, [11] there seems to be silence on how today's tech companies patrol the messages received by vulnerable users. As helpful as such approaches have been, today's media-viewing children and their tech-providing parents are pressured to judge and consume content with intentional and unintentional advertisements in the form of "review" videos, "toy reveal" videos, and candy review videos. [12] These are clearly influential, but the question remains: Should I be concerned about the impact of these videos on children? Is the answer different if the children I am concerned about are less affluent, less supervised, or less connected to positive adults than my own children? Indeed, there are different vulnerabilities for my children than there are for others, and this nuance means that single stories alone are often insufficient in developing practices about a behavior or technology. We need more critical and public discussion about the implications of inequity when it comes to the commercial messages and exploitative labor practices associated with children's TSM experience.

Boys, who are culturally and perhaps neurologically more vulnerable to avoidant attachment and use of aggression, can run a continuum of behavior engagement (internalizing or externalizing negative emotions) and relational compartmentalization (shutting down or avoiding close personal relationships). Girls, who are culturally and perhaps neurologically more vulnerable to preoccupation with their relationships, will overconsume desired attachment communication (nonstop texting) and experience heightened emotional salience of interactions (reading meaning into the rate and pace at which messages are read and replied, for example), impacting offline self-efficacy (can I handle this?), concept (is this me?), and engagement with family and peers (should I share or not?). This disconnect between what is visibly changing rapidly versus the kinds of rapid social and cognitive changes that children these ages are undergoing can leave a child and family system ill-equipped to anticipate the types of TSM exploration that should be associated with these ages. Rather than presuming sexualized behavior within TSM domains is the concern of parents with teens, it's middle childhood where the curiosity, motivation, and interest for information or images is initiated. Just as the toddler learning to walk represents the most vulnerable period for stumbling accidents, late elementary school is the age when children will begin to encounter online stumbling, with some coming from their own curiosity and deceptive skill acquisition, and some coming from the vulnerability of their increased self-directed freedom and the unmediated harm that is easily identified in online spaces, sometimes most specifically in spaces where late elementary age children can be easily found.

Online studies of violence indicate that late elementary age children are readily experiencing online stalking and predatory behaviors; from disguised friend requests to social engineering designed to lure them away from home, these early ages are the doors where early opportunities arise.[13] Yet the skills and capacities that can contribute to TSM exploitation or vulnerability are also times and tools when early achievement is attained. The energy, creativity, and insights of children in late elementary school are valued by producers of content, and children these ages can leverage TSM to achieve incredible prosocial ends, from raising money for charity, starting businesses, or discovering solutions to complex problems.

At the same time, children of these ages, online but otherwise deprived of culturally enriched, community connected, and resource sufficient environments, are at risk for mental health issues, substance misuse, and behavioral challenges that plague adolescents and adults. Deprivation of trust, peer acceptance, social engagement, and cultural tradition erode the resilience supports that are established during these middle childhood years. The failure to cultivate social connection and identity in these ages can be a risk factor for isolation and loneliness, associated with lifetime earnings declines and increased cardiovascular risks.[14]

The reason for these events goes beyond the present moment; nevertheless, our identity forms in large part by late childhood, and early preferences, relation expectations, and our valuation of the expectations we and others have for us has become established by the end of the elementary school years. Popular media certainly convey the poignancy of these years, such as what we watch when child actor Macaulay Culkin's character, Kevin, gets left home alone in the now–holiday classic. While the exaggerated violent antics are likely beyond most nine-year-old's cunning, goal directedness and strategic thinking are all aspects children achieve by late childhood. If English literature is any guide (and many child development specialists will suggest it absolutely should *not* be!), we can note a likely developmental touchpoint for children and families around age nine, when the child's temperamental fit and preferences and their ability to navigate some of their own emotional needs through non-family relationships likely present a vulnerability for premature separation through alienation or entrenched attachment that provides little opportunity for individuation or agency. Joining organizations happens at this age: Girl and Boy Scouts, dance classes, dojo—all of these are identity anchors that are also institution anchors.

Having friends and speaking and playing with them become increasingly sought-after goals, and TSM represents direct channels to peer communication that have always unsteadied parents whose child's bosom buddy is perhaps a little too connected to communication tools, especially when one's child is seeking to prepare for exams. Middle childhood, however, is a space before the transitions of the production and academic performance leaps that are expected in later middle and high school. It represents an important liminal space where capacities for some activities are essentially matched to adults, while the fixedness and self-consciousness of adulthood and adolescence are less present. It is an age when children begin to leave their family circles and engage in their community circles, through lessons for art, life and sales skills with service troops, and spiritual and cultural rituals through religious and cultural education. The child's preparation for the social transition rites of adolescence and the inclinations to vocation and avocation are established.

A child in middle childhood is experiencing a profound set of social and physical transformations, often invisibly and with far less social emphasis and support than might be leveraged for infants or adolescents, the ages and stages when we are often anticipating profound changes.

And so it is, just like with that eighteen-month-old toddling across a highway, children reading to learn are off and running, and children off and running have always been concerning to parents and are cyclically relevant to policymakers when it comes to TSM or other produced media. If I know how few children in my community experience the life my own children do, what should I be doing to be sure they are all as cared for and secure from harm online as I like to think my children are? And what damage might my own device-connected children be exposed to that will influence whether they think about others' needs the way their parents try to?

STORIES, ALWAYS STORIES

We are hardly the first family to fret over our children's concern for others. Psychoanalysts recognized this period of child development as about developing the self-regulatory capacity to appropriately direct libidinal impulses (played out through friends and games) into socially approved and functionally beneficial directions.[15] Pride in winning is to become pride in effort and preparation, sensual soothing is to become intimacy seeking, perfectionism is to yield to practice and patience through increasingly skilled regulatory acts. How have other families navigated the moral and prosocial development needs that assist in achieving prosocially oriented children?

Everything I am today I wanted to be when I was nine. My memories at nine include parties with friends and family, whether that was in a church or in a family home. I loved church, and I was a part of many different ones by the time I was nine, when I was baptized in the pastor's pool. I loved talking with adults about important topics, and I loved watching dramatic television and miniseries, which I did regularly during the height of *Dynasty*, the TV miniseries about dramatic social injustice, and cable TV's nascent mid-1980s spread. I was a voracious reader, and I enjoyed imagining stories and playing with my dog. My very best life these days contains pretty much the same details, with Netflix swapped out for Channel 56, the Boston affiliate that broadcast the syndicated series of *Facts of Life* and *Brady Bunch* I caught religiously after school between second and fourth grades. By fifth grade I had moved to a new household and a new town and moved on to a dramatic one-hour series featuring Michael Landon, who admittedly struck me as the kindly, hardworking, genuine caregiver I imagined my own father might be. I enjoyed a laundry-folding session while blissfully zoned out to the dramatic turns of that day's repeat episode of *Little House on the Prairie*, a household in which children worked hard and life was often unpredictable. It was relatable for many reasons. Even though I am a dramatically different person than I was at nine years old, and the methods for viewing drama series have changed, there is still a consistency about my interests and preferences that was true of myself at nine.

Guides, developmental yearning for mentoring, wisdom: Nine-year-olds in media are famous for their hunger for these figures. At around this age, I was introduced to *Anne of Green Gables*, *The Hobbit*, *The NeverEnding Story*, and, yes, *Star Wars*, released in the summer of my first year, and now celebrated in cringey puns around my own breakfast table on May 4th. Every time and place have theirs (guides and guardians), and these tropes become the figures oft-repeated across a culture's given legends. By the time I was nine, those local legends already included astronauts, and just one town over from mine came a boy who would one day walk on the moon (Alan Shepard, not Neil Armstrong).

Just as devil baby stories played a role in warning families of the dangers of the drink, stories have always been used to transmit values, wisdom, and knowledge to children. Stories function as carriers of culture, connecting children from one generation to the next, to ideas and individuals who can guide their future steps. Yet stories operate in additionally beneficial ways; in transmitting episodic memory to listeners, stories provide a road map of behaviors and solutions that can be drawn from perpetually (so long as the story is remembered or retold). By the time my own children learned about astronauts, it was by way of the Pixar film *Toy Story*, which represents a very different set of values and experiences than the astronaut I came to know. Stories, like that of Buzz Lightyear's fictional home planet, morph over time and place, and so understanding the stories that TSM brings to our children can be a key in understanding the values and solutions that are being presented to them in navigating their world, which is unlike the world in which I first heard, or anyone first created, as that is the point of this stage of human life.

SWITCHES AND SWITCH POINTS

I try to remember this as I navigate children disputing which game to play on the game console or a smaller personal device, such as the Nintendo Switch. It comes into our household just before the pandemic, a tool we decide to bring along on a road trip for a family wedding. In the months before, our family became acquainted with a variety of emerging games thanks to a new position my husband was serving as a game design–lab director. During the pandemic, the house became the remote working game lab, as Oculus Quests and game development software–loaded PCs were prepared for distribution to his students, now remote themselves. They would remain so for another full academic year, and so we became more familiar with the games and tools that the Oculus, Switch, and other game computers provided. It added up, between the available technology in our household and the additional technology from each of our work and service demands, now all relocated into one home, plus our children's school-issued or parent-acquired devices, we likely had more than twenty-one devices registered to our router during some weeks of the pandemic.

Twenty-one hours itself is not an unknown amount of time to be spent in games per week, based on the 2017 Pew study on games and social media.[16] While children in America once attended a moving picture show once or twice per week, children today will engage in stories (via games, television, YouTube, and other devices and modes) for more than twenty hours in a week of moving screens. Of course, that translates into twenty hours less of physical movement per week for contemporary children compared to their great-great-grandparents of one hundred years earlier. It also translates into the kind of techno-socialization and consumer aspirations that support the emerging creative- and intelligence-driven economies in which my children are living. These years are powerful, and the stories in these years are powerful because of the purpose they serve socially for our species and our cultures. No matter the form of screen, the hours used, or the content viewed, we can approach the innovation within this understanding: Our children's mastery of it serves some function and purpose in our social worlds, and if we can recognize this, we can support and co-mediate the experience from a perspective that is both appreciative of innovation and skeptical of its neutrality.

The power in these years and in the tools that allow ever more stories, in ever more varied and impactful modalities, rests in the child's developmental crossroads. While sensitive periods have been difficult to identify holistically for human developmental capacities, there is converging evidence of the evolutionarily adaptive phase that middle childhood represents. During a time when children are not raising their own children, their bodies are able to navigate near-state circumstances in dynamic ways. As cognitive, social, emotional, neurological, and ecological factors are experienced, the epigenetics of the prenatal exposure combine with the contemporary environmental conditions (predictable, safe, resource-available) to result in developmental pathways favoring either a "fast" or "slow" reproductive strategy.[17] In this regard, the gendering that deepens during these phases represents a step in developing a social partnering strategy. The sorting that gendering allows results in competitors and collaborators, and this further contributes to the child's and the culture's advancements, as those skills preferred in that environment are practiced by most members. When culture changes, we see shifts in this age group around the preferences and the expectancies for which they are preparing. As children pursue individual, self-directed TSM engagement, we can anticipate increasingly creative generations, but we should also expect increasingly alienated generations to the extent that TSM is not able to be consumed evenly across generations and groups.

These switch points, moments when development has potential in multiple simultaneous directions and after which pathways are established, happen to children in middle childhood, and they happen in families as they immigrate from one context of their upbringing to a new context in which their children will be raised. Switch points represent multiple sensitive periods of significance because they allow the child, and the culture, establish pathways for development that respond to prior learning, genetic information, prenatal hormone exposure, and material and chronic stress information and results in directed cognitive and behavioral preferences that are adaptive to the perceived and received conditions. It also represents profound moments for support and redirecting generational deprivation developmentally for the child and their family system when supported with the anticipatory guidance that can help the child contextualize interests, urges, and challenges while appreciating strengths, exploring new environments (on city blocks or within Roblox, context depending), and experiencing themselves in new prosocial roles (as sibling helpers or as Scouts, context depending). This adaptive plasticity enables a child's phenotypic variation to be considerable, and this itself promotes a reproductive fitness maximizing life history strategy. While this may help keep culture (by helping a given culture that may have high instability achieve high fertility), it is not necessarily keeping of well-being or prosocial culture, and so support during these years should be recognized as preventing adolescent risks, since the groundwork, social grouping, and social cues for such risks are established in these years.

Where do games and social media fit into this time then, given what we know about children's developmental need for story, competition, and conversation? TSM creates regulatory attachment systems through gamification, and our identity salience drives which system will game us. Marketers rely on this, that what we see in these years influences the path we get on, serving as a flashing signal of early interest and hooking us just prior to our impending awareness of imaginary audience, meaning we engage less self-consciously—unless we are growing up on TV, or on YouTube, like children entering the influencer scene are often now doing. Facebook and gaming platforms like Discord all indicate age minimums, and even the Oculus, which allows VR gameplay but can also allow a viewer to enjoy a 3D movie experience for one, limits its use to above the age of ten (due to visual and vestibular system development risks in younger children).

What I know about children like Ben, who faced grounding if his behavior at school did not improve, was that he had mastered playing his game in the very late hours of his family's home. It was the only time of day that he could consistently access his preferred game, Fortnite, and so it became the reason that he was sleepy and irritable the next day. Attempts to delay or remove controllers always fell short, in part because so many adults commanded Ben that ultimately no one really did. Even if that was not the case, children of Ben's age are able to know others' preferences, and to deceive and delay to obtain personally significant goals. Gaming provides the emotional and cognitive satiety that feels gratifying, and so it creates incentives for this skill to be applied toward TSM. Across all human commercial, technological history, it is likely that this pattern repeats (environment provides TSM cue, parent introduces, child adopts to fulfill communication and competition needs with peers, parents worry over amount of use). The beauty of the pattern is that it can result in further innovation and advancement, and it's worth noting that the game Ben loves to play is a game that involves traveling, learning maps, navigating topography, competing, and, when used with headphones and other players, communicating. A potent blend of a nine-year-old's developmental needs, in virtual form without requiring parent supervision to obtain or arrive at, which children his age would report needing if he were to arrange to meet his friends in the wooded section of the parkland newly recovered near his home.

Over time, my children, and Ben through his game play, will encounter older children and will need to discern their intent as they interact in the game space. These near peer interactions can model for middle-age children how to compete or cooperate. What I was witnessing the morning I contemplated the best way to reply to online hate was my own children navigating together how to compete and cooperate for their mine extraction. Providing affirming, nonexploitive near peer access can assist in the self-regulation and self-soothing practices that assist the older child in regulating for their needs and context. This is something cousins and older siblings have historically represented, but in the context of TSM being competed for in a household or over a hotspot, this breaks down, and parents or other adults need to co-regulate the exchange until the system can in ways that promote the prosocial near peer interactions.

It is a vulnerability, however, that for these near peer children, adolescents, they have not yet developed their own regulatory ability, in part due to the moral reasoning and cognitive development that comes onboard for the child in later adolescence and early adulthood. The child's sense of disgust that enables their contribution to the family meal preparation continues to be absent, but with that absence comes a difficulty in assessing the immediacy of a bad decision. The pit-of-the-stomach feeling that comes from the emotional awareness of disgust at one's own or another's moral action is under development until the mid-twenties, with a range between the late teens and the early twenties. This sense of moral disgust seems linked to the same area of the brain that controls the onset of disgust in terms of food preparation: The insula gyrus increases in activity through adolescence, and it becomes pruned as young adulthood is entered. In Chapter 10, we will explore the influences that act on a child when the insula isn't yet on board, and why the power of a parent's stare can sting for decades.

· *10* ·

Building

$\mathcal{T}$he attachment to caregivers we have in early childhood extends to culture (values, ideals), and attachment to culture results in our attachment to communities. Adolescents develop physically as they increase their capacity to reproduce; they develop cognitively as they increase their metacognitive skills; and they develop socially as they increase their productive capacity, contributing to their home, community, and self-directed goals.

Communities develop adolescents by incentivizing disruptive creativity, which aids in the bonding to community and identity that enables adolescents to build the cultural capital that will allow them to build a life that will lead them into their early adulthood. Where youth build this life, it becomes home, forming their sense of self, tying them to place, and, through that, bringing forward paths and purpose toward which they might not otherwise attach themselves and their goals.

In the United States of the 1920s, only about 30 percent of adolescents did this exploration and commitment in school—the vast majority of youth were at home contributing to the care of family or home, or they were in the labor force—by age twelve for many, and until the age of sixteen by the 1960s. By the time my mother was born, 90 percent of US children would explore and experience their building years inside a school building, and entire systems were created to punish parents and children who did not comport to the schooling requirements. By the end of the 1960s, the investment in US education from the Great Depression would yield a man on the moon, and entirely new battles about which kids from which places would explore and experience in which schools, as communities responded to the bussing that was the integration solution to the *Brown v. Board of Education* decision with protests, riots, and injunctions well past my own multiracial arrival within an otherwise monoracial maternity ward.

Timing is everything, and the consistency that solidifies interests and emerging identity in late childhood develops with time into adolescents' metacognitive capacities. As the routines and habits of self-care and self-direction are tasked during pubertal development, cognitive performance in areas like attention, processing speed, and organizing goal-directed actions begins to solidify for adolescence. How then does TSM influence and direct the cognitive and social growth that forms the basis for metacognition? Various temperamental and environmental factors will ultimately influence the areas of metacognitive strengths and deficits that emerge for a particular youth, but what we understand about metacognition is how it emerges and becomes within greater individual control during and across adolescence and youth development.

Development disrupts. As children's skills, reasoning, abilities, and identities solidify and organize in later youth, there is a great deal of predictable youth and family disorganization that results.[1] This time in a parent and child's attachment is highly vulnerable to conflict, separation, and estrangement, and it is socially meaningful. TSM can be a tool that we use to buffer and repair the ruptures that are inclined to occur, rather than being a catalyst of the ruptures that leave our youth and communities most vulnerable. Realizing this, however, will require our parenting and policy lens to extend beyond our teens' user behavior and into our boardrooms.

MONITOR AND CONSOLE

And so are video games still violent? I don't know. But . . .
realized now that when people are complaining about video games, they're generally only talking about two things. Listen closely: The playing of video games and parenting. That's it. That's it. That's all they're talking about.

—Dr. Kristopher Alexander, Professor of
Video Game Design, Ryerson University[2]

The view on the small thumbnail video I opened in response to the Facebook notification bell showed only a few feet of the courtroom gallery. A teenage boy receiving his acquittal weeps and collapses in relief. Only his and a few onlooker reactions are visible to me, but I recognize him as Kyle Rittenhouse, who, at the age of seventeen, was driven across state lines with a medic kit and an AR-15 to patrol streets of a community reeling from police gun violence against a Black man named Jacob Blake. While there, in testimony and video footage, we saw him at barely seventeen patrolling with the search and secure techniques I see my children use in their first-person shooter (FPS) video games like Call of Duty and Fortnite.

I am not worrying about their use of FPS games, however, as I react to the acquittal, one that strikes fear in me as I consider my children's futures at protests and civic marches. The verdict empowers an ethos we watch horrifically playing out in Michigan within little more than a week after Rittenhouse's acquittal. A fifteen-year-old, with a firearm made in my state and purchased by his mom and dad as a gift, kills four of his Oxford High School classmates. The parents, when told of their son's violent drawings, worrying statements, and ammunition searches, determined that their own assessment of their son's right to a firearm—despite his age, the law, or his developmental needs—were more central to their options than were the needs of the community they were ostensibly a part of. I get the obsession that parents like me had with video games, but when our children face the children of parents who are not even pausing to consider whether military-grade weaponry is required in suburban America, it's hard to pause too long over what it might be doing to them, especially when it seems like non-games are doing something to mothers of sons who lead them to support such unnecessary harms.

Adolescence is akin to the infancy period of physical development, but for emotional development. Where children attached with peer groups and acquired culturally valued skills in early and middle childhood, adolescents are developing their regulation skills. As an infant moves from relying on caregiver regulation to self-regulation in temperature, wakefulness, and activity, a teenager relies on their caregiver to assist in recognizing, responding, and preparing for increasingly complex emotional and occupational demands. Reminder lists, extra trips for forgotten items, last-minute pronouncements of urgently needed items will all trigger memories of these years for many families like mine. These years are challenging for every attachment system; however, the relational attunement, emotion regulation, and attachment behaviors required to effectively navigate the disruptions of development during these years can be greatly trying for some families, some youth, and some spaces. Keys to thriving youth cultures the world over are monitoring behavior and emotional attunement, allowing for responsive interactions that adjust to ruptures brought on by emotion and metacognitive dysregulation predictable during these stages of brain and social development.

Disorganized attachments from abuse or neglect experienced in infancy and early or middle childhood manifest as youth with dysregulation. Making matters even more challenging is the fact that these dysregulations are often happening to the very caregivers who themselves may have experienced abuse or neglect as young, pregnant, and parenting adults: Disorganized youth development in one generation careens into the natural disorganization of this adolescent developmental window and in turn increases emotion dysregulation in the caregiving adult, peaking in midlife.

Over the years, there have been ways that social groups have served to steward youth into adulthood through attachment-organizing practices, aligning youth with identities, occupations, and roles that would allow for some individuation while remaining attached for the monitoring and consolation that the physical risks and emotional wounds of adolescence require. However, youth are not evenly treated within social settings, and many are not in family settings. There is a great deal of meaning that will be made by youth during their adolescence, and what youth make of their youth will reverberate into their midlife and beyond. I found this when I first encountered Dr. Desmond Upton Patton's TEDx talk[3] one day while researching workforce needs for STEM equity. In the talk, he shared the tragic story of youth in his urban education program, students using social media to broadcast the pain, rage, and abandonment they experienced as youth in Chicago. Dr. Patton, a social work researcher, transformed this work into a digital qualitative research career, using social media to understand the ways in which youth rely on social media neighborhoods to keep themselves and their friends safer in their physical neighborhoods. It was also a place where they shouted for help and support—if you could understand the language they adopted within the social media spaces. In 2017, many parents were complaining about kids online, but Dr. Patton's talk was one of the first to show a broad audience how human digital life really was for youth.

LEFT TO THEIR OWN DEVICES

Society often concerns itself with the activities of youth. In the mid-1800s, my state opened its first reform house, for women and youth who were wayward. In a time when paid wages were not guaranteed to children or women, there was not much a youth or single adult woman could manage if ill, mentally unwell, or otherwise unable to sustain work. The only work to be had was a few bends up the river from the home, and they were unable or unwilling to put up with millwork in textile factories; were injured or otherwise excluded due to race, immigration origin point, or unassimilated indigeneity; or cognitive capacity shifts had resulted from neural changes during adolescence. An imaginary audience provides a method by which youth are able to appreciate the social view others may have of their appearance and behavior. As they respond to this, they will adopt the look and preferences associated with the audience relevant to their goals.

Very often those goals have to do with gaining pleasure and avoiding pain. Pleasing caregivers has a special value at this stage, and children are astutely attuned to their parents' emotional signals during these years. In a favorite conference presentation from early in my career, Abigail Baird presented research on her studies of teens. In most research until then, teens were thought not to refine their emotion-recognition skills until later adolescence. In her work, she demonstrated the acute emotion sensitivity teens had, especially in accurately labeling their own parents' emotional signals. Adult faces of people to whom teens had no attachment did not result in emotion recognition, but when an image was flashed of their inscrutable (to researchers) father's face, teens instantly indicated his anger accurately.[4]

It wasn't only parents with whom teens could read emotions accurately; it was their peers as well. In a memorable late 1990s clip from a news report on a shark near a southern state, a set of middle school girls were interviewed and were asked whether they might like to go for a swim now that the shark had been spotted. One girl equivocates, shoots a look to a pretty peer near the center of the lunch table, and indicates she would think about it because in a cage she might be fine. The pretty peer in the middle indicates she would swim without fear, and the nervous girl shakes her head affirmatively.[5] The clip is a boomerang of hilarious adolescent indecision, but Baird argued that it was the perfect display of peer sensitivity, and the socially costly stakes at lunch tables worldwide if you disagreed with the popular peer's opinion.

The challenge both girls had in arriving at the safe answer isn't only due to a sensitivity to peer affirmation, but a not-yet-developed insula, the parts of the brain and stomach that communicate gut reactions of disgust and guilt. Watching an antihero in a drama perform the irredeemable act that you know will have devastating moral injury on him and others provides most adult viewers of villain stories a pit-of-the-stomach response. For teens, few will report this, especially in their middle-school years. Anything is possible when you don't immediately register these strong, morally associated, physical sensations, and the insula's late "onboarding" in early adulthood shifts youth into adulthood by organizing behavior.

When I was twelve or thirteen, I was grounded for more than eight months. I don't recall what it was that prompted the first round of being grounded, and what I remember of those years mostly involves rerun episodes of Michael Landon TV dramas and Tracy Chapman "Fast Car" music videos on MTV. What I also remember is that those two things frequently kept me from taking out the frozen chicken in time for dinner, which makes me think my grounding had something to do with that combination of memories.

In any case, grounding involved not being able to attend after-school activities, make plans with friends, or enjoy extra time with TV or music outside of my room. Sometime during this eight-month grounding episode, a general testified about some Iran-Contra affairs, and a baby fell into a Texas well, and all I know is that I recall both of those events vividly because it was the first time in each case that I had gotten to watch TV in weeks.

What my mother believed would come from my grounding, I can't say. At the time we were surviving a domestic abuser, and yet we were living together in as calm a condition as we would for the remaining decade, and I was thriving in school. What was my mother attempting to regulate by placing me on a two-week rotation of social and media restrictions? What did eight months (or maybe more) result in for her, or for me? Ultimately what resulted was our being alive and home together when she ultimately decided to leave that partner and move with me and a younger sibling to another state. But when that happened, it was a year or more later, we lived in a different apartment unit in the building (one where, incidentally, the state's first "battered woman" had killed her own abuser only years earlier), I was in a different grade altogether, and whatever cycle of grounding and reprieve that existed had stopped being relevant as I assumed greater household responsibility. Within those abrupt shifts in what would have been only fifteen months, my worlds changed drastically. This same period for many adolescents represents the developmental period of puberty, and with it, the social-niche leap that adolescents experience as a result of their physical changes. The environmental response to those changes results in greater social responsibilities and expectations for social behavior. The tensions experienced and expressed between my mother and my twelve- to fourteen-year-old self are hardly unique or new, and, by most accounts, the version of these events today would also have included restrictions on game consoles, online access, and social media use. Deprivation, it seemed to my mother at the time, was the key to helping a child regulate their leisure time. Regulating leisure time would result in regulated non-leisure time, which would result in more dinners, made on time.

While it's easy to judge this harsh parenting, well, harshly, it's also easy to understand the conditions in which a violence-prone home can reduce episodes of violence by having consistent mealtimes so the violence-prone diabetic in the household doesn't experience a life-threatening (his and ours) episode. If my lackadaisical TV viewing made it more likely for our home to erupt in violence or death, then I guess extremely long (and apparently not quite effective) deprivation seemed mild. Regulating the environment, at least the distractions of it, was easier than dealing with what happened if there wasn't this regulation.

Of course, having a parent or caregiver home after school, a set of circumstances that did not involve hourly labor exploitation or limited social services for a family, would also have made a difference in regulating the conditions of our home at that time. What was the best answer? Readers reviewing this portion of this book might rightly disparage the approach my mother took and the conditions we were subjected to as a result. I have my own judgments, although knowing the circumstances of her life and the options available to women in those years, which are even less in some ways than women now, I understand the choices she navigated. Ideas as to what else could have been done—about the home, my distractedness, or her vulnerability—are many, and none are necessary.

Today, we have similar perspectives when we read headlines of parents distracted by tech while their child suffers a fatal injury, or headlines of children unable to engage in basic classroom activities because of too much screentime. We look to the environment of the child, note the parental behavior, and recognize that regulating the parent would have altered the outcome for the child. We are primed for the individual actor, and our solutions are frequently created with that individual in mind.

For a few years when my children were still very small, I worked to help develop a program to support families impacted by homelessness. Collaborating with regional child development experts, we sought to identify a program that could support the parenting, limit-setting, and redirection skills from which the parents of children of any age benefit. The program we selected[6] was modeled after the idea that specific emotion-regulation skills combined with stories of flexible families and characters would enhance a family system's capacity to roll with the hard times while still being able to experience joy and establish limits. It was a time just before ubiquitous smartphones, when the families we were serving were using trac phones and the agency answering machine to leave messages for their case managers. We were not solving for social media overuse in children or parents at that point in time, but I knew from my own TV media–obsessed tween years that the older kids we served were likely dealing with similar family demands with vulnerable younger siblings, and that all of them were dealing with the emotion dysregulation inherent in houselessness, family separation, and addiction recovery. I wanted the program to bring in adolescents for some of the parenting lessons we were delivering; however, this never came to be.

What I had recognized in advocating for teens to receive the same lessons as their parents was that in many lower-income communities, it would be tweens and adolescents who were tasked with much of the limit-setting, flexibility, and adaptability within a family system. Just as tween-age children enter a distinctly valuable evolutionary niche, adolescents navigate a social niche that is shaped and prescribed by their social contexts and historical time. For the adolescents being raised by the middle-class white women who would implement the program we designed, adolescence was a period of protected vulnerability,[7] when adult responsibilities are forestalled and activities are socially oriented with minimal accountability.

But the adolescent I had been, and those we were serving, were precociously mature, having developed hyperarousal, tenacity, and authority due to the earlier traumas that separated them from and were faced by their parents. While judgment and decision-making may not always have been mature or in their future interests, the adolescent children served in the program were often treated socially as more mature. I recognized the benefit of their being supported in healthy media engagement and limit-setting with young children, whose charges the younger children were highly likely to be left within as parental use of program supports grew less intense with prolonged housing stability.

As adults' lives stabilized with housing, treatment, and family reunification; their social safety nets would begin to fall away, and at the same time, the older children's stability through housing, reunification, and newly formed school connection meant adolescent children served as the social net their single parent (still recovering from the cognitive and emotional damage of trauma, addiction, and chronic stress inflammation) needed to rely upon as they sought to juggle the erratic school, medical appointment, or daycare schedules that would threaten their newly earned hourly at-will employment. There was just no world in which a family would not end up relying on a modestly responsible preteen, and every likelihood that would extend into their under-supervised early adolescence.

Often enough, the program would soon become the treatment and housing program for the children—who themselves would become parents or addicted as a result of the under-supervised precarity forced on their family systems by a state safety network that incentivized the underemployment of middle-aged adults receiving state aid for families over the supervision and family support needs of adolescent children of the same precarious working-class workforce. Maybe there was a way that children learning about how engaging media can be, could be used to support the skills in their younger siblings and model for parents the practices that promote healthy media, and ultimately healthy family engagement.

Instead, what youth engage with TSM is often bonding them to communities and identities that allow them to conduct the tasks of this developmental period within close enough circles to make the kind of emotionally exaggerated risks of youth with and around, while allowing enough exploration for the productive social engagement skills to be solidified. The additional tasks associated with these ages and stages include the task of establishing one's method of how they will contribute to their social group and in what way they will take and direct their earlier curiosity, reflection, activity, or connection and engage in their greater world. This building of one's identity through engaged action is the crux of experiential learning approaches, methods much celebrated when engaging youth. While the maturity possible in adolescence offers some benefits for a family system that has care demands, their maturity also results in increased risk as children become subjected to harassment, stalking, or assault.

Adolescence is a time when being away from adults has benefits and risks, and the ways in which TSM enters the relationships teens and families have within themselves and between others is significant. Individual families, regulatory agencies, and corporate makers of the TSM itself cannot navigate the reality of this paradox well: There is universal agreement that TSM's problem centers on the impact and issues it raises as a result of children's and adolescent TSM adoption across platforms and the adult pursuit of them on those platforms.

This was something else we learned, then, which is that TSM shapes and is shaped, but that shaping only extends so far. What this meant was that while adolescence might have been recognized as a time of risk and vulnerability for many physical threats, TSM as a location appeared the most protected place for our wanderlust teens to explore. Maybe they might see some dirty pictures or deal with some strangers, but the fiber they journeyed over was a far safer place than the asphalt sidewalks and empty (non-adult-supervised) apartments that my friends and I had traversed in our teens. Or at least, that's how it seemed.

It wasn't always like this for adolescents; the model of adolescence for centuries was that capable young people would apprentice in some community-level professional work, master the entry-level skills of that work, and conduct their further vocational development by working at their socially sanctioned professions. Certainly, it wasn't a sure way to make every scullery maid a Michelin-starred chef, but it was a model that contributed to the social utility of an age with great capacity and considerable inconsistency in performance. Today's youth using TSM are operating globally in several different, competing economic realities, but most of those realities[8] emphasize adolescence as a stage of life without status or rights; engaging in professional exploration to contribute socially to one's group then becomes fraught for youth who have limited embodied knowledge of their non-TSM worlds,[9] while absorbing the cultural messages about their cohort and the social expectations of them. In this context, is it no wonder the ways in which youth whose futures are foreclosed by limited community expectations and vocational exploration are such volatile challenges for their families and communities? When the cultural reality of youth lives meet the sociolegal realities of opportunity offline, TSM becomes an even more central environment, where exploration, including vocational, will necessarily play out.[10]

An important technology touchpoint for teens then is to recognize that TSM will serve an economic function for those whose status and preteen experiences will orient toward financial support of themselves or their families. It helps us orient our attention to the reasons a child might begin selling 3D-printed guns,[11] or how children might find themselves rulers of licensed products and YouTube empires.[12] In large and small ways, adolescent TSM engagement opens economic doors for youth that contain unregulated risks (labor exploitation and performing, for example), and itself because of its interconnectivity to currencies of various sorts means that TSM will be rewarding (providing financial and social support in the form of paralinguistic online emojis, credits within a specific game, or a streaming platform economy), but with a specific sets of rewards. Their relevance for a teen's ability to bond with meaningful others and build toward an adulthood that protects them from bonds they reject while binding them to identities with positive status is critical.

SCHOOLING AND SCROLLING

Before he was a professor, Kristopher Alexander was an esports athlete, ranking top in the world in two different games and at times averaging fourteen hours of play a day to maintain his performance well enough for the global competitions. He is emphatic that video games may be consuming, and indeed too consuming for children needing to master educational material and engage in professional preparation. However, he's equally clear that Facebook is as much of a "game" as any game he has trophied in.[13] Scrolling through Facebook can add up to the equivalent of seventeen football fields in distance scrolled. And when that scrolling happens, whether on Facebook or another TSM, we are increasingly scrolling in gardens organized by race, class, and gender—even if we perceive it otherwise. Gaming community researcher Kishonna Gray writes:

> Although the era of public segregation may be gone, modern digital segregation mirrors the historical practice of designating space as whites only. These practices come in many forms, including lack of inclusion, toxic environments, and outright hostility, harassment, and violence in many contexts. Virtual spaces are direct mirrors of historical segregation, as overt racism permeates them.[14]

Professor Kristopher Alexander's respect for the self-actualizing potential for TSM and human self-regulation and goal-directed behavior is admirable. I consider the meaning of this scrolling, knowing that it was the scrolling and the communities that Rittenhouse's mother and he were scrolling that brought them (and their AR-15) to Kenosha's streets, and it was likely the same TSM feeds that fed Jennifer Crumbley's TSM accounts, where she shared about the mom and son day at the shooting range that prepared her child for his massacre.[15] What is the likely effect of parent feeds promoting social separatism, disdain for authority, and mistrust of helping professions during a global trauma and in the context of poorly regulated physical environments? It seems there are significant effects, even as these aspects that contribute to a cultural value system of armed young white men as civil servants of safety and independence proceeds without commentary. When context online mismatches context offline, as it did for the youth that launched Dr. Patton's work on digital neighborhoods and gun violence, and resources are uneven in contexts, as it is in the communities where Rittenhouse and Crumbley violently connected, then alienation and the salience of that isolation increase. It's a cycle ignited easily by the power of the TSM systematically crafted for such ends. We will be less concerned with global deterioration when our youth are consumed with the facilitated destruction the tools enable. The emotional hopelessness delivered via TSM streams increases individual sensitivity to stigmas, leading to increased withdrawal, or nihilistic engagement.

Adolescence is a dynamically powerful biological and social transformation, when the young of our species suddenly become the adult of our species, without the burdens of the adult loyalties that stifle community change. This is a developmental, social, and psychological set of ages, and the impact then of the TSM now present for youth is one that cannot be ignored. The way in which all of this itself is bound together strikes me, and I recognize the TSM anchor that once again appears in the national conversation parents navigate around their family's use of tech. I turn to my TSM outlets of choice to share an eloquent observation. I am heralded in the TSM manner–likes and care emojis. In *Intersectional Tech*, Kishonna Gray writes of how the solidarity formed of storytelling can be leveraged to challenge, deconstruct, reconstruct, and renew, even in the face of great trauma or vulnerability.[16]

Kishonna Gray conducted a years-long immersive study of/with Black female gamers online. Gray writes compellingly about the specific traditions, technologies, and techniques that Black women gamers have shaped through TSM,[17] and this work anchors the remainder of the book in so far as the lessons and techniques Gray outlines are implicated as we try to trace developmental realities against technological potentialities. Little recognized by parents whose caregiving has turned them out of the youth culture ready for their youth, this is the power of TSM for teens, as it allows them—like Black women—to share and shape stories, while not risking the physical and physiologically overwhelming *in vivo* experiences outside their capacity, affordability, or accessibility. Despite this, sadly, lower-income, BIPOC, and femme-identifying children face conflict no matter their context, online and off. It's these realities that result in the oral and storytelling knowledge-sharing traditions that TSM has been able to convert into cultural currency, if not actual control, for Black and adolescent cultural production online.

There are implications of the power of TSM when it combines with our innate neural circuitry and when it confronts the time- and experience-bound limitations of human development. In building one's sense of self, while not in the three-dimensional local world of one's physical body, how does one come to know these things about themselves? How might not knowing about these things impact one's options later in adolescence or adulthood? There was little protection for me when as a young teen I was knocking on neighborhood doors to raise money to help my family by selling raffle tickets, and there would have been little options for my family had I not been collecting the weekly forty dollars my efforts managed to supplement the meager earnings of my mother, a hotel chambermaid. Many of the parent comments on the technology parenting groups point to the lack of supervision that feels possible as our increasingly adult-appearing children navigate their online existences. At the same time, much of the research about the health needs of adolescents emphasizes the power of supervision, monitoring friends and providing support to engage in community. It certainly feels impossible to know when our teenager's internal awareness of how they interact online will kick in for them, but there are developmental touchpoints that, of course, are underlying these patterns too.

LOOKS AND LIKES

As children approach adolescence, their sensitivity to humiliation and rejection increases. The cultural values and rules that had been absorbed and mastered in childhood now become judging standards that are not attainable by them at the specific time when they are most sensitive to social feedback. While parents can feel at odds in providing supervision for online activities, there is a little monitor on the shoulder of every teen in these years as their imaginary audience serves to alert them and orient them to achieving socially relevant goals and actions. Their judgment will be formed by our own parenting earlier in their lives and the unique cultural reality their generational cohort is navigating through its TSM exchanges. Sometimes we get peeks of what that process looks like for our own children. I was struck during an interview with a podcasting philosopher who in his doctoral work is exploring transhumanism and the ethics of online living. When we explore the techno environments of our own emerging adolescent selves, he shares a story I recognize immediately from my own late nineties computing existence.

In Seth Villegas's case, it was a scene of his family room circa the early 2000s, Windows 98 on a gray-body Dell computer. He was there to chat and play games online with friends from across the street and the world. And when he was there, he was there, fully present to the online interactions before him. The moment that began changing his enthusiastic relationship with TSM was not from an imaginary audience, but from an actual one, one that in its physical salience would immediately redirect his engagement as more internal and less expressed when using the family computer situated in its most public space. Seth, a PhD candidate at BU and podcast host for the Center for Mind and Culture, focuses in his work on transhumanism (the ethics of humans physically incorporating computing technologies into their physical bodies). Although we discuss the ethical dilemmas inherent in the rapidly growing transhumanist effort, when I inquire what some of his earliest TSM memories are and how they shaped and informed his work today, he is a good sport as he recounts the memory. Even through the Zoom screen, I feel the chagrin he experienced as he describes when, while gaming online with friends, he realized his reactions and expressions were all witnessed by not only his parents, but also by two of their friends, friends he had not previously met. The embarrassment was palpable.

> And I was so engrossed in the experience that I just didn't . . . I didn't know that they were there, right? I, I didn't know anyone was home even, right? Until I was sort of stopped. And that was actually one of those things of like, "Oh," like, "Maybe I've taken this too far," um, right? Like, uh, maybe you know . . . And even like, even the way I'm behaving isn't something that I'm very proud of, right? It's something that I'm embarrassed to have other people see. So, maybe there's something wrong with that.[18]

Seth describes navigating an emerging sense of his online and offline selves and negotiating ways to display and engage that match the context of his actual life while meeting the fulfillment and belonging that comes from online engagement for him. He is extremely conscious, though, that part of his continued ability and interest in doing that same thing, even as an adult beginning his professional career, came from the bonds and boundaries prepared for him by his family and cultivated and committed to in his own actions and interactions. When the pandemic shutdown in the spring of 2020 returns him to some of the online games and spaces his intensive college and graduate studies had moved him away from, he is struck by the acute self-consciousness experienced and projected onto him as he exited spaces in his reemergence from the pandemic lockdown. What was a place to visit again was home still for many. When that offline aspect is mismatched to the online fulfillment, there is a degradation of those whose offline existence is full. In that rejection is a desire for validation, affirmation that existence of any kind can have purpose and meaning; the snarky comments as he logs off to reconnect with in-person contacts now give him a self-conscious sense there is an online audience feeling rejected by his engagement. When we talk, I can't help but recall the Thanksgiving Eve bar nights of my hometown, and the ways in which local kids who stayed local experienced themselves, as kids who went away and visited during college and grad school break returned. I did both between those years, and I felt the same palpable cultural and status shifts accruing from education and TSM engagement, offline parallels to Seth's online experience. Greenfield's work is about these patterns within a given time and place, and the way TSM cultivates the "society" conditions of Gesellschaft.

The developmental period of youth encompasses a variety of individual and social group developmental touchpoints; as parents approach their most socially burdened years, their youth are entering the world in ways that are like and not like the parents who are already out in the world. The degree to which youth align with parental goals as a normative social process will often depend on the degree to which resources, status, and wellness are available to the caregiver adults in their lives or to the ways in which their children's economic options mirror the options in their coming-of-age years. This has been true from across time; and in the modern era, this has implications for what types of guardrails are possible or probable for youth.

Today TSM integrates but segregates social experiences between and within adolescent communities. As children's cognitive and emotional developmental tasks of mentalizing (taking on other perspectives and using that to reframe one's own understanding) and regulating (emotion regulation) evolve over these periods, they work to bring the adolescent into contact with a sense of their future self and make the pathway to that self available. These ingredients will turn motivation into actions by sustaining a sense of purpose, regulating the emotions needed to pursue and attain it, and orient toward the behaviors and knowledge that serves the purpose, which through behavior creates the conditions where the developmental tasks of intimacy and identity can be established. TSM creates a landscape that may be artificially segregated, organized, and reflective of power inequities that foreclose specific opportunities for youth engaging TSM.

The inattentiveness to the real risks online for children became itself a revelation and a movement launcher. We learned in the fall of 2021 via Senate testimony that those children in the" becoming" years, particularly nine- to eleven-year-olds, were the emerging target audience for apps like TikTok, which captured children and their household networks and the biometric data therein.[19] Why the interest in such detailed households when seismic global shifts are so clearly underway? Information, particularly such linked and surreptitiously received information, is powerful, and youth often have reason to challenge state controls. Is it any wonder this would be the health effect that motivated the Senate hearings (that of the health of our national defense) and the risks of TSM as sources of state and global conflict while appearing to our children as entertainment channels?

We are facing socially the largest population of adolescents over the next decade globally than ever before. This means things for societies experiencing inequity, conflict, and crisis, all of which are increasing on our horizon. I was feeling pulled by the vulnerability of my children, each emerging from the ongoing pandemic trauma in their own ways, each simply living their lives and engaging in the toy and peer connection opportunities of their moments. The spaces where their exploration was unfolding were being identified as places where their sense of self and opportunity was being manipulated, and as such, the health opportunities of their engagement were affected by the economic conditions of the non-TSM realities they have. And yet it is also the case that TSM facilitates purpose and meaning exploration and can with care and intent be maximized to achieve Adolescents 2030[20] goals to increase global adolescent well-being. By 2030, the world will see its largest generation of adolescents, while undergoing its most serious climate change and greatest human migration due to climate and geopolitical realities. The ingredients for bonding and building preteens' connections to TSM, enabling the harnessing of this global youth power in our society, are primed. After the demoralizing headlines in Kenosha and Oxford, I find myself doubting whether we are as ready as we need to be. In Chapter 11, we examine whether the adult mediation of TSM, and their uses of TSM to fulfill adult development needs, might serve to expand our conceptions of TSM safety and innovation.

• 11 •

Bridging

$\mathcal{T}$o hear David Blanchflower[1] tell it, I have two years before I am the unhappiest of my life. Given that the past two years have been consumed by a pandemic and the remote schooling of five children in four different schools with eighteen different teachers, I have to admit I am a little horrified by what's to come for me mood-wise. The foundation of his claim is that across 132 countries, over both recent and longer-term trends, 47.2 years of age is the lowest mood point for humans. Living in my father's native Caribbean islands might push that up to 48.2, but either way, his data suggest I am staring down a tunnel of despair, although the fact my father is a native of the Caribbean and is Afro-Caribbean itself suggests my despair may not be so low compared to the Euro-descended side of my family. White people, it seems, are particularly despairing in his review of US data across the post–Great Recession years. We are all living that despair, deprivation, and developmental touchpoint every day since, thanks in part to the power of TSM and its function within the stage of human life known as "adulthood."

Adulthood can be thought to have rather squishy age bounds. Socially or legally, adulthood could be reached at the age of fourteen, sixteen, eighteen, twenty-one, twenty-three, twenty-five, or even thirty-five, all ages that have certain legal and social shifts associated with reaching them and our obligations or opportunities to engage civically and economically within the broader group. While adolescence is about bonding to a group and building a sense of self, purpose, and a path, adulthood is about journeying along that path to self while finding meaning and fulfillment in our purpose. If we can, we become generative, contributing something of social and personal value that will outlive

135

our specific lifespan. Certainly, a generative contribution of our adulthood might be even more ephemeral, but to meet the developmental tasks of adulthood, it will need to mean something. TSM has an interesting way of shaping adults and their adulthoods. It shapes our sense of self and the sense of meaning that makes up a life, shifting our expectations and influencing our social engagement and our pursuit of experiences.

Everyone has a tech story. In Chapter 10, I reflected on the power of the imaginary, and actual, audience that greeted Seth Villegas when he shared the memory of his parents and their friends watching him play a video game online in the late 1990s. While Dr. Desmond Patton's story is different (and one we will return to in later chapters), I am struck at how each of the stories these two men—whose current careers focus on the ethics and safety of TSM—share a common factor: For both men, technology in their lives came through the adults in their lives. A family decision to purchase a home computer (for Seth) and a father's career in computing (for Desmond) are the bridges that each man journeys across on their own road to technology. Rather than his father's computing programming career being a career to which Dr. Patton was drawn, the fact of his father's career simply made the technology landscape of his own life that much more technology-enriched and -enabled. Adults bridge worlds between one generation and the next, and often it's in simply living their adult lives that TSM enters the world they create with their own families. Such was Dr. Patton's experience; while his story didn't involve building PCs or internet companies from the ground floor up, it was the enrichment that brought him to appreciate the ways in which TSM connected him to communities, and how that connection revealed needs and gaps that youth used TSM to address.

This is a critical observation, because it isn't simply markets and moon shots that drive TSM innovation; or rather, those influences alone don't spread TSM into a given child's hands. Unless and until an adult provides access to TSM, a child's facility with it will flounder. Children generally cannot procure and sustain independent access to TSM (at least in its early-adopter phase, where expensive hardware and fee-driven internet access is a reality). Youth innovate TSM, but adults institutionalize and inculcate technology, providing a contact zone for technology. In this contact zone, they introduce the cultural and social forces[2] that technology can mobilize or modify through individual user engagement. Adults bridge their youth culture to emerging youth culture through modeling and making technology available, and selecting that technology comes through the motives that adults are meeting in their own tech use.

MOTIVES, MEANING, AND MATTERING

Generativity is the task of adulthood, or at least that is what Erik Erikson argued as he, in his own midlife phase of development, articulated his foundational theory of ego development. As a student of psychoanalysis, he was interested in understanding and outlining the ways in which humans across their lifespan develop through and from our earliest dependencies into reflective, contributing adults. He argued that adulthood was characterized by building a legacy through one's communal actions; whether that legacy was a family or a career was less significant than the pursuits, and purpose, that humans derived from the efforts. The meaning we derive from our experiences may shift over time, and indeed, when writing about later life development as an older adult himself, Erikson acknowledged the ways in which shifts in meaning for our own past can result in despairing older years.[3] Even so, for human adults, our activities between the cultural age of majority and our own end-of-life period are filled with self-selected and culturally determined pursuits.

These pursuits are not only for mere survival, on which adults are necessarily focused, but also for personal accomplishment, relationships, and status. By midlife (an undefined time that can range from as young as one's thirties to as late as one's fifties), adults are developmentally focused on making sense of their lives and making decisions about their futures, but for most midlife adults, the options for shifting their pursuits or realizing a sense of meaning from them will be shaped by the early adulthood years. The reason is that one's early adult life, extended as it is from the culturally ephemeral adolescent stage, will be tied to the actions and relationships anchored during one's "defining decade."[4] Without a motivation to attend to others' interests and needs, and a matching desire to then meet or exceed in these areas, there would be little that might explain human feats of engineering and civilization. Human developmental needs of generativity are functional through increasing humans' engagement in culturally valued tasks and deepening one's relationships at a time when less dependence on others might otherwise be needed.

The evolutionary function of these shifts across adulthood serve to organize behaviors toward developmental tasks, thus shaping personality and
role identity in short- and long-term ways. Without the psychological
rewards associated with, say, raising children or writing a book, there
would be little that justifies or sustains such activities, and yet these kinds
of labor-intensive, minimal immediate-return investments do orient action and desires. We are motivated to pursue what we believe matters
and what makes us feel we matter,[5] and that pursuit provides adulthood
with profound meaning.

Goals,[6] the internally derived and socially valanced ways in which
humans motivate their and others' behavior, are the substance that makes
up the adult fabric of life. Goals shift according to our own perceived
capacities, competencies, and in accordance with external realities.
As younger adults, we tend to value growth-oriented goals, deriving
personal meaning through the relationships and activities we perceive
will assist us in reaching our full potential as culturally conceived and
communicated through status. In middle adulthood, we are oriented
toward goals that will consolidate our growth and contribute to future
generational success, even if that future is not one in which our own
children will live.

The goals of middle adulthood are more often pro-socially oriented
and in the service of individual aims to understand prior life difficulties
while navigating later life with fewer goal-related challenges. By older
adulthood, motivation and goals again shift, orienting behaviors toward
retaining health and material gains made in early and middle adulthood,
and in the service of maintaining independence and ability as we navigate the losses associated with aging, from cognitive fluency to motor
abilities. Across all chapters of adulthood, mattering, deriving meaning,
and motivation orient us to value social relationships, and the formation and retention of them remain highly valued goals for humans, even
while the specific relations may change according to age and role status.
Their importance to our sense of well-being is consistent. We use these
social relationships to construct our own goals and rate our capacities for
attaining these goals through our relational connectedness.

In this context then, we might appreciate the despair that Blanchflower's[7] research indicates I am facing and that I argue we are all living
through in our community and TSM engagement, but what explains
this creeping despair that threatens my family's well-being and, increas-

ingly, our democracy? As my developmental needs for consolidated achievement and personal meaning reach a peak at midlife, I am simultaneously experiencing shifts in my social relationships. Dependent children are increasingly independent in actions, while the financial and psychological realities of their dependence deepen; physical and cognitive abilities previously relied upon and predictable are now subject to shifts (of disks in my spine or lenses in my eyes). As these realities converge, my own sense of self and social connection also shifts. In the midst of these changes, I have less cultural mentoring and care than would have been available to me as a younger adult. In middle age, the availability of senior mentors is highly contingent on my circumstance and the road maps of those whom I might most directly encounter, encounters increasingly gender-, racial-, and age-segregated and tied to social and occupational roles adopted by me earlier in my life. I see the role I hold in my family shifting at a time when personal meaning and mattering are important in determining my goals and life course.

BURNING BRIDGES

Similarly, in the years following the Great Recession, those whose social positions had always been constitutionally secured (white, male, educated) and the people attached to them and their social status (namely, white women and youth) began to slip in status; first in household wealth, income, and educational attainment, and more and more in terms of social relevance and racial dominance.[8] The diminished social and economic status (relative to dominance and highly unequal generational wealth by race) combined with increased (by location) demographic diversity, economic inequality, and social degradation of liberalism became a potent resource exploitable[9] by Trump and those who saw an opportunity for capital and religious ideological gains through his dominance and authoritarianism. The fact that the insurrectionists on January 6, 2021, were entrepreneurial, college-educated (including post-graduate degrees), employed, and in middle age is a significant one;[10] and the fact that they were called (and collected) using TSM[11] is not coincidence,[12] considering media framing and TSM-increased status salience of white TSM users. Media manipulates through metrics, and when numbers alone drive programing and messages, it is easy to understand how micro-targeted, social-status relevant messages could be used to reach the adults with the leisure time, assets, and identity-driven interest in preserving a specific frame of a lifeway.

I am on Facebook and Twitter but not Instagram and Snapchat, because my skills in creating words, rather than my skill in capturing images, drive my interest in words from others. As features that made one platform popular with specific subsets of users, they would appear in TSM (Twitpic absorbed by Twitter, Facebook acquiring Instagram, both eventually incorporating the platform and technology into each other); this is not unique to TSM, because broadcasting media have always cultivated audiences; it is why advertising works as a revenue stream for media companies. As an audience, adult use of TSM is likely not a singular set of behaviors and is not singularly motivated by our desire for social connection; rather, these reveal a likelihood that my and others' TSM adoption and use will be driven by the personally meaningful goals I have directed myself toward. We need to take risks that are socially valuable and personally relevant, or our sense of self will suffer, and being an aging human complicates this need. Adults lose social connections and resources over time, in part due to self-selected behaviors and in part due to circumstances in our social ecology, each of which interact with each other to further shape or constrain opportunities as we age.

To cope with these shifts, humans adjust their goals. For centuries, the weight, measure, and opportunity of socially valued, personally relevant goal-setting and attainment in the individualized sense we use them were greatly constrained by our social roles and status—a far greater predictor of life outcomes than our personal interests or task performance. Some economists argue that this is still true and plays out in the context of societies like the United States and Europe, where debt ratios and social debts are unpayable, with foreboding consequences.[13] The local group and economic form would determine the local person's life goals, both in the fact of its resources to support their pursuit and in its support of lives that deliver some type of community resources. Now TSM extends our sense of "group" beyond our local in-person realities; it bridges our local in-person reality to global places and realities, and in doing so, it shapes goals and, increasingly, economies.

In the course of our lives, these variations are referred to as social clock projects.[14] Social clock projects differ by gender, generation, and generativity: Conformers seek lives that adhere to dominant culturally sanctioned scripts; achievers seek lives that fulfill individualized achievement goals, but also provide recognition of those achievements; and seekers are defined as pursuing lives that resist social convention and are marked by an unsettled pursuit of expanding awareness and self.[15] In this context, TSM alters generativity aims and outputs and influences our own social clocks. Our engagement with TSM is intergenerationally impactful, and the way TSM brings generations together is little appreciated but results in transmission (of culture), transformation (of socially relevant information), and transfers (of wealth, ideas, and other high-value material) by providing methods of identifying, establishing, and attaining goals not only with youth but across the whole of adulthood. That adulthood is about meaning is perhaps a surprising nonbiological need that otherwise directs actions, but it should not be. During my interview with game professor Kristohper Alexander, we explore the ways in which emerging TSM and the adoption of it within school, work, and homes has impacted opportunities for those previously excluded from community. He shares:

> If you have a phone, you can broadcast yourself right now. But it's the message, it's who you're talking to, it's the community that you serve that is going to determine where you go, right?
>
> —Dr. Kristopher Alexander, Professor of
> Video Game Design, Ryerson University[16]

His observation, that TSM individuals are empowered to share themselves and their ideas and through that sharing can find community, is one that resonates, and yet its simplicity glosses over the implications for our current moment. From the perspective of a woman of color who functions online, and further from the point of view of an American in 2021—I have some misgivings about the power to present my message and have it spread by community, especially after speaking to community is what landed me on a hate list. Not all bridges of communication are sound, after all. As the novelty of discovering our individual capacities to establish community wanes with aging, the need to be motivated to continue to engage in prosocial and otherwise voluntary actions would have to increase to see the kind of collaborative, cooperative, and coordinated actions that make running a household, a company, or a community possible. What we now recognize about TSM is the way it has been strategically deployed and economically capitalized upon to burn social and institutional bridges that would otherwise contribute to socially generative outcomes.

While the disruption of youth has a societal function, the despair of adults has a social impact. The January 6th insurrection had as much to do with despair as it did with white supremacy, and the violence instigated was not by youth but by middle-aged adults navigating generativity versus stagnation, and among the older rioters, ego despair versus ego integrity.[17] Far from being only individually impactful, despair has societal impacts.[18] This spreading and affirming of despair made possible via TSM is less appreciated and represents a special risk given the post-pandemic time, the rising youth population, and the global economic and climate shifts occurring: Tomorrow's millennials are very quickly becoming middle-aged adults who will face similar developmental pushes, as well as crises of conscience or faith, as all despairing adults do, and have to reach the other side with their resources and relationships intact to weather the storm of aging that follows those blessed to live beyond age forty-seven. TSM introduces entirely new opportunities to discover our capacities.

The data that economist David Blanchflower shares about midlife despair are not definitive; the research is evolving, and it's possible that happiness and life satisfaction levels that increase into later adulthood are due not to perspective taking and gratitude but are due to people surviving to the degree of their individual optimism. But it's important in this social technology touchpoint to consider that as desperation to reach identity and status anchored life-goals peaks—and as its associated images and affectations transmit across TSM—attention and actual personal and family resources are impacted. When this happens in the context of social times when economic resources are stretched, when families are vulnerable to conflict, and when social optimism is lower, there is bound to be a special vulnerability to the escapism and escalation that TSM enables,[19] and pandemic circumstances certainly increased adult engagement with TSM during times that were challenging to mental health for its economic, interpersonal, and well-being impacts. It's a predictable risk of TSM that financial risk-taking, asset vulnerability, and demands on personal, interpersonal, and social resources reduce social resilience due to generational impacts.

Technology-empowering TSM was made for and by white men, and this has been true for so long that collective amnesia has formed[20] about the women—Black, Asian, Latinx, Indigenous, Middle Eastern, and, yes, European—who have ultimately made TSM innovation possible. Writer Kishona Gray comments that

> being embedded in the core of technologies, whiteness is defended and continues to be replicated and reproduced, unseen and unacknowledged. We must explore design and structural bias, creating the core of technological artifacts.[21]

Even before such geopolitical machinations, however, women online in the mid-2010s were becoming increasingly aware that these free-market spaces, where thoughts and ideas could contribute to fraternity, liberty, and egalite, were anything but when they were in the hands of those motivated to deny entry or engagement. Between 2014 and 2015, specific female journalists, and then many women online, became targets of coordinated smear and harassment campaigns, all designed to expel voices critiquing the increasingly socially material video game industry.

It was not the case that women across racial lines were informed of how bots worked or what types of public statements could disempower abusers; instead, white and Asian women navigated the attacks often on their own, a shocking revelation for me as someone who witnessed and attempted in small ways to deflect the harassment as it unfolded. It was an interview with Emily Robin Dreyfuss in late 2021 when we were exploring the technology bridges traveled in her career as a technology journalist, writer, and researcher. As a beat reporter and then-editor for *Wired* from 2006 through 2015, she was less aware of the interconnections between nefarious, socially motivated actors and the geopolitical implications of their effects. For me as a mother and woman of color online during those same years, I was all too aware of how the TSM threads were being weaved to shape public opinions and promote a certain kind of policy solution. It's an observation confirmed in Kishonna Gray's work:

> The revisionist construction of white victimhood has led to digital and physical harassment and attacks, threatening of women and people of color who dared to venture into the "Whiteman only" space of video games. White masculine anxiety has become increasingly legible for a larger body politic since the election of Donald Trump. We witnessed a very pronounced iteration of it during

the Brett Kavanaugh Supreme Court confirmation hearings. Senator Lindsey Graham (republican South Carolina) and Kavanaugh
unleashed the white man's backlash in visceral ways, and Graham's
statement captures the essence of this backlash: "I'm a single white
male from South Carolina, and I'm told I should just shut up, but I
will not shut up."[22]

In building bridges in online spaces between 2010 and 2020, it became
increasingly obvious that online space would be battled for just like in-
situ space: Power holders (whether it be only social power associated
with one's race or gender, or actual power associated with one's reach
and influence) would not yield without contest. TSM allows such con-
tests for power to be impactful in ways few reporting and shaping it at
the time appreciated. It would not be until 2015, after the trial balloons
that led to "yourslipisshowing" safety campaigns, and after "gamergate"
had resulted in harm for journalists Brianna Wu and Anita Sarkeesian,
that technology journalism would begin to appreciate the confluence of
risks and controls playing out within the evolving TSM spheres.

Dreyfuss reflects with me what 2015 represented in terms of TSM:

> That was the year we realized algorithms entrenched bias because al-
> gorithms were created by human systems. It was also the year that we
> realized that scale itself is . . . Can weaponize things that otherwise
> are not dangerous—and it was the year that we realized that there
> was never going to be politics without technology. There was never
> going to be parenting without technology.[23]

She continues:

> The harmful network effects that are enabled by the infrastructure
> of social media as it currently exists. There are so many harms and
> a— and, and . . . And maybe not necessarily intentional harms.
> Like, well-meaning actions that happen on these networks that then
> because of the way they, that social media is structured scale up to
> create these real-world impacts extremely quickly, and in a way that
> is very hard to recognize, very hard to separate yourself from, and
> extremely difficult to fight back against.[24]

What might be taken for the kind of professional awareness that drives
industry-wide insights must be understood as evolving from knowledge

that must always be considered but is easily ignored in the context of TSM. Gray writes:

> To make sense of the interactive relationship between user and technology, it is necessary to understand power, social hierarchies, historical legacies, access, and a host of other concerns that make up the asymmetrical power structure of digital technologies. This asymmetry is based on an unequal distribution of economic and cultural resources, which are closely linked. This perspective reveals that there is no end goal when it comes to one's relationship with technology; rather, this is an interactive, ongoing, negotiated process whereby one continually influences the other.[25]

Families left to their own devices as they watched a far-right insurrection unfold in Washington, DC, on January 6, 2021, were becoming privy to its power, as hundreds—spurred on by the narcissistic and nihilist leadership of Trump—descended on the Capitol building. TSM was being deployed and helped along by malicious foreign and domestic actors seeking to trigger white nationalist fantasies of race riots that would result in civil war. What I saw in the days leading into January 6th, extending as a line from #yourslipisshowing through and past the hate list incident had occurred to me in summer of 2018, was the kind of scale and spread that Emily spoke about, and a lack of awareness and attention to the power dynamics that Gray's work so eloquently shared.

At the on-the-campus talk I attended at the beginning of our fall 2018 term, Stanford education professor Sam Wineburg reveals his research on countering misinformation. He shared his 2017 research on how experienced fact-checkers dealt with online information relative to novice users. During the talk, the auditorium screen lit up with the websites that his novice and expert fact-checkers were tasked to examine. As he progresses a slide, he shares from his studies, and I discern in the margins of the website a signal that suggests to me a white religious separatist group is behind the content, which is being passed off as a pediatrician organization promoting child well-being, when it is really a front for preventing LGBTQIA+ affirming care for youth. In the disinformation research Wineburg presents, there was no mention of that signal of separatist hate; rather, his research presentation was suggested to provide an emphasis on the "lateral" search procedures experts in his study used to verify sources of a website's info. It worried me greatly that while fact-checking was being considered an essential skill set that we ought to promote with our students, there didn't seem to be much awareness of the identities of those who might be motivated to avoid such treatments of their material within TSM spaces. Wasn't a digitally literate college youth population one that could easily identify ideological agendas in information or organization mission? What Wineburg was attempting to heighten in college students through his information-literacy pedagogy was a complex metacognitive task, one that can be cultivated through practice and rehearsal, as his method allowed. It is also, however, a developmentally timed capacity that emerges forcefully across adolescence, with significant risks for children and varied risks for adults. It is also a cultural task, since detecting and deciding on facts to check will be driven by prior experience.

Testimony from TSM employees, founders, and whistleblowers has trickled and reached a crescendo as the evolution of technology continues. With TSM, how an adult could engage it to personally valued ends varied within my lifetime and generation. As emerging adults using the nascent Web 1.0, today's middle-aged adults were then teens, connecting to bulletin boards, consuming information and cocreating it in small networks. In our professional adulthoods, Web 2.0 has watched us move our most personal digital information from a CPU stationed near our feet to a cloud we can only conceptually identify as it allows our creating and curating. It is our hands now that create and plan for the emergent Web 3.0,[26] where increased local control is imagined with fewer buffers between TSM users and those who would defraud them. Like the middle-age phase might be thought of as about controlling and collecting, so, too, are the latest iterations of technology and human flourishing. While a foreign government known for violating the human rights of ethnic minorities having my household data and biometrics is abstractly terrifying (abstract because there is no condition I can imagine where such information will matter to my daily life), I know I am naïve not to be alarmed by that reality. I am also aware that what changes I have seen were ones that seemed at the time jovial, and youth-powered: When Trump rallies during the pandemic leading into the 2020 election were sold out only to be empty, it was not due to some foreign actor but rather due to US TikTok videos encouraging users to gobble up and then not use the Trump rally online ticket registrations. I find the weight of these implications to be more than slightly overwhelming. It might be vestiges of Red Scares in Senate testimony rooms that makes me question the motives of political leaders, but I can't deny that what lawmakers during Senate subcommittee testimony are asking CEOs of major platforms about, the degree to which they track and monitor youth users and the degree to which other foreign governments or parties might have access and control over it, are legitimate questions.

In reviewing the Senate testimony more than a year after those seemingly innocuous incidents, it's chilling to recognize how manipulable such events are in the uncontrolled feed of TikTok. In some ways, each trend and topic is one that we shaped in our engagement: How does it reflect on our family's evaluation of the trends? What was the value we gave and got from adopting it for our home? It is a question of value and values, one that will be the focus of Chapter 12.

· *12* ·

Value and Values

*E*very technology comes into being because it enables something of value: The value may be in what is *made possible*, or the value may be in what is *made*, but in either—and powerfully when both are true—that which is seen as creating value (in terms of tech) will be valued, and therefore more of it will likely be enabled by additional value, perceived or imagined. In similar ways, the economy of attention in our home or day-to-day interactions with others reflect what we value, and our attention to that which reflects what we value deepens that attention and value commitment. In navigating TSM, we must consider both how its shape reflects what is valued in a market sense, but also in the moral sense, since it is this interaction that feeds the dystopian nightmare we battle today.

If we wanted to visualize a straight line between prior intersections of humans and now, we might have an image of a Don Draperesque–ad man holding aloft his VR headset-wearing child on his shoulders while waving a cigarette lighter in his hand and a lit-up iPhone in the child's displaying a "lighter app."[1] Capturing the scene would be a cathode-ray camera circa studios of the 1960s, and the family would be viewing a live-stream of a concert connected to a laptop with YouTube. Maybe that is too much for one image to capture and communicate, but those details represent both the transformation of the media values and the valuation landscape, as well as the regulatory reality for habit-forming products, of which TSM is but one. Humans are and (re)make systems, we tend to only draw from a few design models when we do, and what we do with those systems takes similar shapes and patterns as the things we made with those designs before, no matter how innovative the thing is that we see as new technology. When researchers and regulators alerted to the harmful effects of cigarettes and the marketing techniques they used, Americans began to see the lengths high-value corporations would take to protect the value of their stock above the values of care and respect that adults wanted to believe drove leaders.

Values and valuation change our relationships to technology, each other, and our environment. When we change our relationship with or through technology, we are always altering a natural process and introducing a different relationship than was possible prior. These relationships are not neutral and renegotiating or reframing them has impact. Anna Lowenhaupt Tsing, in her work,[2] refers to the ecologies of human interruption as feral, where life springs forth in patterns and patches shaped not exclusively by human action, but through the relational changes human actions make possible and the "world making"[3] processes of all earthly organisms. Using humans' use of fire in forestry and marginalized communities' occupational and social resilience skills, she articulates a complex story of biological change and human capital that—in micro—mirrors the macro and proposes ancient and affirming ways humans might adapt their relational practices (among and between governments, corporations, human communities, and ecological niches) to survive the Anthropocene.[4]

Rivers are one type of ecological community. When used in the colonial Americas, rivers were bidirectional, but human cognitive and social relationships to rivers became unidirectional during the Industrial Revolution.[5] This fact limits our solutions now as we seek to address the continued pollution that we now know will linger in our soil and cells for centuries. Many environmental cleanup and restoration projects now being proposed are all about a return to our earlier bidirectional relationship with such places.[6] At this powerful intersection of values and valuation, we have seen both what is possible and what is potential for ways in which these types of intersections have been navigated socially in our recent national past. Our awareness of these forces and their possible outcomes is useful in our consideration of how, what, and whether to move forward or against the next cycles of TSM barely ahead, and certainly already shaped by what we have enabled. What does that look like as we navigate rivers of technology that have heralded our information revolution, powered by the investment of its profits in ways not unlike the industrialists before them?

At the end of the century, a very young boy left home to find his fortune. He was briefly educated, quickly apprenticed, and within a short time he was able to turn his attentiveness, surety, and insight into world-changing finance, transportation, and telecommunication networks the world had never before seen. I could be describing Alexander Hamilton[7] (in the late 1700s, and the Federal Reserve), Travis Kalanick (in the late 1900s with Uber),[8] or Mark Zuckerberg (in the late 1900s with Facebook),[9] all of whose biographies contain the pluck of poor or working-class young men motivated to remake the world, and who despite all odds, become successful beyond even their imaginings in doing so. I am, however, referring to Jay Gould, in the late 1800s and his railroad and telegraph, and financial, empires. His rail and telegraph networks ran parallel to or were connective of the Indigenous waterways the American expansion project of the mid-nineteenth century had turned into unidirectional forces, and he did this building immigrant and newly emancipated workforces (and nascent unregulated government agents) to extract natural resources cheaply while boosting settler European commerce. For that, Jay Gould amassed a destabilizing fortune, and he transformed America (the land) and America (its wealth and identity) permanently. In his lifetime, he was pilloried in the TSM of the day, the printed newspapers of New York, yet the world he built to connect commerce and markets would be the very foundations that would later build our own transformative commercial and communication infrastructures.

Globalized workforce-management practices fueled by imperial conquest result in this ethos being spread, both in Gould's time and in ours. What could drive us to reevaluate our TSM relations? Embedding ethics into technology development and marketization could look a lot of ways, but at present it mostly looks like reactionary corporate resets. Uber—a transportation logistics app that seeks to transform itself into a riderless car company—has wreaked havoc on local urban and immigrant transportation services (themselves regulated by strict local operating rules) while making "no lines, no waiting" private transport possible for the well-heeled and young. Its founder was disgraced by allegations of poor impulse control, lax dealing, and hostilities toward Uber drivers and women.[10] Despite such patterns of behavior and reputation, however, many seemed willing to proceed, until the deafening crescendo of dissent grew from shareholders eager to save their own investment. Even with a change in leadership, though,[11] the new male CEO reported he felt no particular need to review the independent investigative report authored by Attorney General Eric Holder once it was completed; he suggested that while the aggressively masculine "deals, deals, deals" culture might have contributed to mistreatment, it led to growth, so that should be the focus he keeps in the forefront as he pursued a reset.[12]

In short order, the next Uber scandal headline came and demonstrated how not knowing our history, personal or corporate, limits our understanding of the future[13] and revealed how bound to global financial capital America's technology boom actually was. In Uber, whether with Kalanick or his successor, Dara Khosrowshahi, little examination would be done to understand the ways in which TSM appears on valuation sheets and how that appearance manipulates the humane values that might otherwise present. It's these weights—and the associated interlocking, self-replicating, and blocking these weights provide in terms of lobby, monopoly, and meta-versal potentialities—that truly drive the morality, or lack of it, possible in TSM platforms, devices, and user-policy statements.

In India, long a country where creative ingenuity as an ethos (*Jugaad)*[14] is required, that same value system is being blamed for broken financial and social contracts, with harmful effects at scales much larger than a single ingenious problem-solver fixing their tractor with homemade parts. When the *Jugaad* or the *"move fast and break things"* ethos is applied in TSM development and deployment, its impact can't be minimized, and the consequences for millions can't be ignored. Techno-solutionism has altered human values in favor of private valuation.

TSM promised connectivity, opportunity, and an unboundedness to realities, but adult social experiences of TSM playing out over the past half decade reveal that with such promised opportunities came actual disconnection. Increasingly families discuss divisions caused by TSM use; adults describe harassment and fraud visited upon them through the porous doors of TSM; and workers in industries vital to the protection of democracy or the oversight of TSM describe the myriad ways surveillance, disinformation, and harm are visited upon them and progressive causes that might otherwise improve the lives we all live with and through. Ellen Ullman has written more eloquently than I on these matters. In her memoirs of emerging technology and shifting women's status and roles within technology companies, we see the degree to which measured, iterative, and thoughtful design, program, and release cycles have been abandoned for rapid, monetized deployment and shaped by individuals whose value systems were built on ideas of individualism that do not sustain systems at scale.[15]

Decades later, with systems of capital shifted because of this enduring ethos, it is routine for early releases to be both monetized and announced as "in beta" (as in testing phase 1 or 2). This means that even though a user might perceive an app or other TSM innovation as secure and predictable, in truth, consumers cannot take for granted that careful assessment has occurred when they adopt a technology innovation, but rather they should presume that *they* (and their children) are the test—even for products and platforms years old. TSM innovation, development, and deployment is often not applied with care, and therefore the values that promote and sustain it are not restorative.

Patricia Greenfield in her work explores capitalism and social development. Greenfield explains that in any market, capital can displace, and values within the existing capital good system will always pull toward opposing disruption, since capital functions best when there is predictability.[16] The answer isn't to void all TSM innovation, but it is to appreciate, attend, consider, and define how TSM and the technology enabled by it might be better constrained in the service of promoting humane values of dignity, self-determination, and collaboration.

I know that not everything established through the values and valuation of TSM is terrible. Around my city, I can trace the Industrial Revolution and the periods of reform that the Gilded Age ushered in and washed away. I can appreciate the journey of a young entrepreneurial adolescent[17] striking out on his own after seeing the violence and community challenges wrought by economic dependency. I, myself, left home young to make my fortune (my education actually) possible, but, in my case, it was to live in a rooming house built for farm girls who once toiled in factories and on telegraph lines fed and fueled by Gould and others, selling raffle tickets door to door in neighborhoods I never imagined I would afford to live. Maybe it wasn't the safest choice, but I got to know about myself, about my strengths, about my challenges, and about who was in my corner.

REPAIRING THE BREACH

The antidote of despair is hope; and in attachment theory, the antidote to rupture is repair. I turned to my online TSM community ever more after the hate list in the summer of 2018, and I cheered as figures long known or newly known made statements and took stands of their own against encroaching hate. Online, I connected with an organization working to recruit mental health professionals and community healers in order to support a march for Black women to encourage passage of the reauthorization of the Violence Against Women Act, which was set to expire (and did. It's reauthorization took four years more, signed by the president who had sponsored the original legislation when he served as senator in the 1990s). I had previously trained as a disaster psychologist, so I found the chance for short-term service near people I already enjoyed visiting—too well aligned to pass up.

It was on the Brooklyn Bridge that I first encountered the limitations of my hope, and someone who inspired it, while trying in vain to make the DC version of a march song I learned a day before catch on with the New York marching group (it didn't, though I can't sing). Humans, and the systems they create, are imperfect. To hold in one's conscious mind an appreciation for the best parts of the world that they are navigating, as well as an awareness of its shortcomings and threats, adults hold both in their minds, because doing so allows us to continue moving forward in achieving individual and group goals. In community, adults must trust others enough to believe they will honor their intentions and not harm or take advantage of them. And yet sometimes, that trust is misplaced, as even Gould would learn in his life,[18] and we have to be aware of patterns and trends that might indicate mistreatment is on the horizon. Humans in community, whether virtual or not, must learn from mistakes and be optimistic enough to keep making them. Hope is a society's prenatal vitamin.

While I was moved to support others like myself marching for Black women affected by domestic and intimate partner violence, I knew that online violence was an ever-present reality and that serving at the marches was a way of reclaiming hope from the fear and alienation being placed on the hate list had brought. As I prepared to travel, I prepared materials for my psychology students, who were reading work on culture cycles by UC Berkeley psychologist Helen Markus. On the drive down to DC and New York, I found myself focused on the cycles that have influenced and shaped TSM over my and my children's navigation of it.

Markus and Conner's 2014 book, *Clash!: How to Thrive in a Multicultural World*, introduced the concept of a "culture cycle" in which *ideas* shape *institutions*, *interactions* therein, and *individuals*—within and outside—and the model explains that influence can move in any direction; a striking speech by a striking person can influence individuals to commit actions, and those actions can result in institution or be institutionalized. In *Clash!* riveting stories of how geological, social, and economic practices form culture and in that shape values; in communities where attachments to land meant ancestors favored rice cultivation, women's workforce participation is high; small hand tools and loose soil meant women could farm and harvest as well or more easily than men. Likewise, in areas where agriculture required heavy equipment, in animals or machines, women's workforce participation is lower; the values of gender egalitarianism can be understood as having roots in—well, roots—and how easily space could be created for them.[19]

While these geographical influences on culture were fascinating, it didn't mean women were not getting beaten if rice was the primary crop four hundred years ago. What is true for countries must be true for every part of the ecological system surrounding a given child or adult TSM user. How might a technology ethics framework apply to the ideas, institutions, interactions, and individuals that make up a TSM's culture cycles and the clash of values they sometimes raise? What would a technology ethics frame provide as ideas to shape institutions and the interactions that have such profound impact on individuals and the earth's resources, but especially for some people and some resource ecologies? What happens when individual people from various backgrounds and communities are not present to interact, influence, or spread ideas? Perhaps the lack of responsiveness within Twitter and Facebook to earlier alerts from marginalized voices hints at what happens and portends some solutions.

Announcements in the spring of 2019 stemming from the Department of Justice investigation conducted by Robert Mueller[20] reveal that several TSM network policies and vulnerabilities within both Facebook and Twitter were intentionally exploited by Russian foreign actors in 2015 and 2016 to spearhead a massive disinformation and community acrimony campaign. In the revelation, it becomes clear that practices (fake profiles) and policies (reporting abusive or denigrating posts) became tactics for false accounts to target or attract followers of certain social media accounts. While congressional testimony and reporter filings suggest the unexpected nature and freshness of these concerns, Black women in 2014 were alerting Twitter and each other to the existence of these manufactured campaigns, yet nothing was done by Twitter to address that user community's concerns; rather, nothing from the company was done.

Online, Black women themselves rallied to share the info[21] and protect others by starting a hashtag campaign and engineering block lists to minimize the time demands of having to individually assess and block hundreds of thousands of fake accounts. They also taught each other how to easily recognize such accounts (improving user literacy in the process) that would want to follow women of color with opinions about social justice. It was a social media literacy lesson sorely needed by more white Twitter users, but sadly Twitter's inaction made sure the Russian bot program to sow racial animus and disinformation about the 2016 election was a success. Interactions (online) can indeed shape institutions.

While the American democracy suffered, Twitter benefited from increased users and accounts, which from quarter to quarter impressed their investors enough to keep the share price despite a consistent lack of profits since its founding. The wisdom of the Black women Twitter users in creating safe spaces and calling out harmful actors extended from the kind of community care Black women writers have often taken up for their own and others' well-being. Generative aims of similar protection and collective healing formed the 1971 Combahee River Collective, whose architects, Barbara Smith and Audre Lorde, had argued these principles were required for Black women. In my Twitter feed, I began seeing women whose contemporary actions were reminding me of that revolutionary power of presence and fortitude in the face of malevolent forces. Women like Bree Newsome—who scaled a flagpole under threat of electrocution to remove the confederate flag from the South Carolina capital after the Mother Emmanuel massacre—and women like Therese Patricia Okoumou—whom I watched climb the Statue of Liberty to protest family separation just months before the march.

What are the dangers we are navigating in our current TSM state, and what do the advances already being commercialized and considered mean for the metacognitive capacities for tomorrow's humans? The evidence—that self-learning occurs through experience and the value of embodied knowledge that one develops through experience and reflection is essential to human development—is increasingly clear. It's why facilitating access to technology and technical skills for youth is a consistent workforce-development strategy. The opportunity TSM provides as a portal to the skills and jobs of the future we are building makes it a public utility in terms of the significance for the well-being of a given citizen, and youth are citizens whose full enfranchisement rests on the continuous access to opportunities needed to sustain concerted effort without devastating interruption due to deprivation, victimization, or criminalization. Our identities can limit how aware we are of the system of which we are a part.

STORIES AND SCAFFOLDS

To live, one needs hope, and serving at the marches for Black women as a healer in that small way restored mine. Stories do this, too, and songs, and I got filled with both at the marches that fall weekend. The images of women I related to on Twitter were made real and especially poignant as I passed a statue I remembered raising pennies to help restore, then myself a child separated from her parents (the reason for Okoumon's Statue of Liberty protest was to draw attention to the children in captivity at the southern border).

TSM allows humans to share stories of redemption, deprivation, victimization, and crime, and we are drawn to human dramatic content. We write our own dramas in many ways across TSM platforms. James Pennebaker is a psychologist who studies narrative and physical health. His research shows that even thirty minutes of emotion-focused writing can reduce physical illness over three months. Everyone thinks the next frontier is external in energy and technology; yet the essential frontier is internal. It's how we see, respond, value, and grow everyone with purpose and pride. Feeling disconnected (not understood, known, or able to be fully seen), feeling refilled on a regular basis (through individual, community, and spiritual activities), and recognizing successes as they unfold will make it easier to sustain our connections, though TSM can challenge us in its ability to do this. We are not infinite energy generators. We are social creatures, and we operate within social structures that rely on emotional attachments.

While TSM may feel liminal, status and social offenses in the actual world can easily become felonies in the virtual one, even as it's also true that sweet, angsty posts might get noticed and result in money for college or a business; the potentialities are potent when it comes to

TSM and human development. Where do we turn for answers? What can regulate such an unwieldy system? In the United States in the early twenty-first century, teens became adults as much offline as they did online, and the adulthood adoptions and mediation of their TSM at the time was less about the care and responsive attachment (and internal audience) than it was about how commercial computer salesmen conceived of creating a market for adults to use computers not only in an office, but in their everyday lives. I find this in a *New York Times* magazine story about growing up in Palo Alto, California, circa 1999. When I rediscover the article, and remember my own 1999 visit to Palo Alto for graduate school interviews, I realize in 2021 how the family PC that Seth Villegas fondly recalled in the Chapter 10 interview wasn't placed in his parents' living room because of their good judgment about where it should be located; rather, it was because a company in the 1980s had hired researchers to study ten families and how they engaged when at home; since the living room was where people were social, it became a place the researchers identified where a home computer would receive the most use and engagement, and might allow for other technological adoptions in the space outside the kitchen with the most family use. Markets are made by less-thought-out strategies, but it's clear that our current market for TSM has its origins in this evidence-based approach to TSM development.

What Seth recalls as his family's planning filters into his sense of the degree to which he was supported in overall ways. The degree to which his engagement was supported while also allowing the kind of careful self-exploration and reflection possible is bonded to both his own individual capacities and to those of the systems of relationships around him, built primarily through their attachments to one another, to him, and to their community and the identity they played in the various spaces they inhabited. Little recognized by parents whose caregiving has turned them out of the youth culture ready for their youth, this is the power of TSM for teens, as it allows them—like Black women—to share and shape stories, while not risking the physical and physiologically overwhelming *in vivo* experiences outside our capacity, affordability, or accessibility. Despite this, sadly, lower-income, BIPOC, and femme-identifying children face conflict no matter their context, online and off. It's these realities that result in the oral and storytelling knowledge-sharing traditions that TSM has been able to convert into cultural currency, if not actual control, for Black or adolescent cultural production online. There are implications of this when it combines with our innate neural circuitry and when it confronts the time- and experience-bound limitations of human development. Manipulations of attention and relational attunement, which TSM so powerfully does, are easily manipulated economically, socially, and politically. Examining TSM for adoption then requires consideration of the social and technical impacts of those whose economic and social experiences are less privileged, as writers critiquing the opacity and limited capacities of machine learning and artificial intelligence increasingly call for us to do. In *Fake AI*, writers Parida and Ashok argue this applies not only to developing children, but for nations:

> Especially for countries in the nascent stages of AI development, it is important to scrutinize proposed solutions with a socio-technical lens and critically evaluate the potential of power imbalances and foreseeable positive and negative consequences.[22]

It shouldn't surprise us that industries created by technology developed out of and for military surveillance, coordination, and control have the same power and impact in our homes and schools, nor should it be surprising when challenges to TSM hegemony are framed in consumer interests, but are opposed only to the extent they represent strategic interests. Still, it surprises me. In the library book sale, I came across a volume of encyclopedias on China. Seven hardbound books featuring hundreds of pages focused on China in the post-TSM world come home with me in the reusable grocery bag. As I begin volume one, page after page features testimony from 2011, just months after my youngest child is born, and the same year that graphene is applied to microchips, increasing computational power by more than sixfold.[23] The testimony focuses on the uncompetitive internet technology business practices adopted by China and negatively impacting human rights and well-being. The testimony is framed as focused on the human-rights implications of China's "Great Internet Firewall," where access to country servers is controlled by single fiber line routing. I prepare to read testimony that will feature details about the Uighur Muslim minority in China, reportedly confined to work camps and systematically alienated from mainstream social participation by the controlling government. What the testimony emphasizes instead, however, is the negative impact of the firewall on US business viability, with great emphasis on the future costs of being shut out of China's five billion internet user market.

The calls in the 2011 testimony reveal the economic shortfall that was happening and being forecasted. This, combined with testimony about the explicit antidemocratic effects of internet control approaches in China, made it clear that in 2011 the arguments against TSM platforms and their potential for harm were not really about individual growth and development, but about the economic and resource access fights between nations; in this context, individual human potential will never outweigh the pulls of national interests in a social world order that continues to demonstrate the resilience of systems collaboratively connected while the power of systems competitively connected drives the innovation approaches used to navigate crises. What does an approach to TSM development and deployment look like that considers both the captain of industry's interest and the multiplicities of captains and the interests they could have? What kind of technology approach listens to women, or children, or parents, or the elderly, or the community impacted by the mineral or labor extraction required to sustain it? Our flawed attachments to organizations lead to degradation and devaluation of humans, and we are wise to consider the forces and factors that have shaped the current landscape, even as the horizon presents such dramatic departures ahead.

Understanding the act of creation and the way in which reality is shaped and perception is formed by the TSM structures and processes is a vital cognitive leap that is too easily hidden from parent TSM adopters and policymakers. Understanding processes and ways of increasing or decreasing the likelihood of certain behaviors that we think are associated with resilience—things like cognitive flexibility, faith, friendship, and habits of health—would empower users in ways current approaches focused on capturing attention have not been able to achieve. So far, approaches at institutionalizing such awareness and shifting practices are weighed on the shoulders of adult users who build the bridge between TSM and their households through purchase, services, or promoted use. Technologists and civics experts are beginning to demand more. As TSM scales ever larger, companies imagine controls that will meet challenges to access and choice. But the proposed solutions once again obscure user choice and control. As AR and ML solutions to abuse are trialed, there is nothing that ensures their fundamental rights of humans in the march to a collective future. It's in late 2021 that the United Nations makes its demand to halt the government use of AI and ML until such safety and considerations can be addressed. Even with these calls, adoption has increased.

I speak with Dr. Desmond Patton, director of Columbia University's SAFE lab, late in 2021 via Zoom.[24] We spent a few years collaborating on a grant to support youth STEM education, a continuous effort by our government to increase interest and capacity in youth like the ones on which Dr. Patton's work has focused: poor, Black, and with limited exposure to technology since their working-class parents were often not the ones making tech. He founded SAFE lab to apply social-work principles to technology study and application in real-world communities. As his career began, he valued the stories youth provided on Twitter and other TSM spaces as clues to the community trauma they were navigating, alone but with others in their "digital" neighborhoods, the places they would often turn when their own urban neighborhoods neglected by policymakers were unsafe. His work was among the first to suggest that TSM posts themselves were useful markers of community need, if only the humans reading them understood the youth codes being used to express their fear, rage, pain, and sadness. When I shared what congressional subcommittees were hearing about TSM in 2011, he was not surprised. Acknowledging the gap between what technologists and teenagers in community need, he shared:

> Families did not care about social media policy, or tech policy; and not in the way in which it's currently articulated in popular media. And it's something that actually I have a big issue within tech policy spaces, because what I've experienced, what I continue to experience as someone who's often brought in as a consultant or a working group member, is that the conversation is not one that's grounded in the lives of diverse people. It's really about, you know, who has power here and who gets money there, who has control here. It's not about how people actually experience their lives through tech. And how that experience with technology offers different insights around how tech policy and technology should be developed in a way that actually helps them as well. It's not about powerful people, but about everyone. So that's what I expect from some family members. They do care about tech policy, they care about their kids! They want their children to be alive. The want their children to have maximum opportunity. They want their children to have opportuni-

ties to aspire and to dream and to build connections and community. They want all of the fruits and opportunities that tech can bring. But they are encountering choices and decisions that were not made by them or for them. That then filter and you know, disadvantage their young people in ways that they can't control either. And now we know that. Many social media platforms have been optimizing the most aggressive experiences, and this is particularly challenging. You live in a community where violence is a part of daily life, right? And so optimizing aggression does not, you know, bode well for folks who are living in communities with high rates of gun violence. That is just chaos. And so I think that what parents want is for a tech-lived experience that benefits—again, people—and also keeps them safe.[25]

I have been walking through these hazards and holograms with my own multiple hats, but these mirror the rings of influence that are at play. Urie Bronfenbrenner conceptualized the rings of a child's life as concentric circles enveloping the child in family, community, society, time, and place.[26] So it is with TSM and any notion of individual relationship with it; no child is an island, nor is any tech user; while not matched in ability or impact of individual influence on the circles of systems around either a child or me, there is always an interplay between these. What are the possibilities of that influence, the limitations of it, or the consequences beyond my individual use or child's access? These are the bigger questions that TSM demands, and too often these are not the questions able to be navigated well within the TSM spaces themselves or in the public square that has historically been where humans have sought some level of shared governance and engagement.

• *13* •

Regulation, but Whose?

*I*t is understandable that discussions of TSM and children will center the child and their parents in terms of regulating their use and access. This same centering of the child and their family is a key foundation in child development theories as well. The circles of influence around children that Russian American psychologist Urie Bronfenbrenner outlined in his bioecological theory of child development includes the family, neighborhood, culture, society, and historical time. This observation revolutionized the ways in which community support was conceived for children; attention was paid not only to household circumstances, but also to increased awareness of how the broader community might be a resource to support children's development. In 1964, Urie Bronfenbrenner, by then a professor at Cornell University,[1] testified before the Senate and helped persuade the Senators to support early community-based resources for families of young children. As a result, Head Start became a life-changing resource for the poorest four-year-old American children, offering meals, early exposure to school practices, along with play-based learning. Decades later, this social effort would be borne out with higher educational attainment and better lifetime health. By the end of that same decade, Fred Rogers testified in support of public television, sharing insights into how children's learning and emotional well-being would be enhanced with national, commercial-free, educational programming available to every family with a television set.[2]

The 1960s in the United States was an era of both great social turmoil and great social programs, and it was the turmoil, and the coalitions it created, that ushered in the era of change. Public programs such as Head Start and PBS, and departments such as the Environmental Protection

Agency and the Occupational Health and Safety Association, were shaped as part of the Great Society era of reform, as political pushes for meeting social needs through applying social science and government resources garnered the public's attention and support. With the Supreme Court decision of *Brown v. Topeka Board of Education* in 1964, and with the passage of the 1965 Hart Celler Act, opening immigration from Caribbean and other countries, this was when classrooms and communities were becoming racially integrated.

As the 1960s wore on, the United States was moving to enshrine, albeit temporarily, protections for voters, workers, children, and the environment. It was a time when the government had a role to play in ensuring the well-being of Americans, not only through defense and social safety net programs as it had in the past. Of course, there was hardly uniform support for these efforts; many interests coalesced to prevent such agencies from forming or from programs being funded, but they formed and flourished despite these efforts. There were still many, however, who strongly felt the government wasn't needed to interfere into how you treated workers, your land, or your children. As integration efforts and impacts unfolded, combined with the inflation of the early 1970s and a reduction of organized labor's power happened in northern manufacturing cities, the political support for expanded social programs and integration began to shift.[3]

Well into my 1980s elementary school years, white parent protests around the bussing of minority students from the suburbs of Boston played regularly on my local news. My community didn't struggle with these conflicts; I was one of three children of color in my elementary school in the mid-1980s, and my younger siblings would enter the same district a decade later and still be some of only a handful of minority children well into the 1990s. The northern cities didn't require Jim Crow to keep their communities segregated[4]; they could rely on limited community support to deter those not already well-established from joining the community.

The three eras of reform in the twentieth-century United States are known as the Progressive Era, the New Deal, and the Great Society. Robert Britt Horwitz, in his examination of the telecommunications industry shift from monopolies to deregulated entities, describes in *The Irony of Regulatory Reform* these reforms. In the 1900s the Progressive Era came on the heels of the Gilded Age, which Gould and others had established, and it was about protecting the interests of small businesses (and investors) in competition. The Sherman Act, passed in 1890, ushered in the beginning of antitrust law in the United States and became the first to regulate rail and the emerging telecommunications infrastructures, such as they were in the telegraph years.

The late 1800s and early 1900s constituted an era when increased immigration and poor housing conditions, and the abuses of industrial workers in cities, ushered in protests for worker protections and a push for community services. While labor-related social changes would grant increasing power to management, some protections were established. In communities, the social upheaval and community concern for controlling immigrant children resulted in federal social support agencies in the form of the Children's Bureau; there, to be sure, the devil babies of Addams's lore grew up with some proper attention. Worker protections unfolded; women's hours and children's labor would be limited—a patriarchal compromise for the overall worker protections that were being agitated for in the streets following workplace tragedies[5]—however, most laboring men saw few worker protections from these reforms. Businesses were subjected to anticompetitive limits; the Clayton Act in 1914 and the establishment of the Federal Trade Commission and Department of Labor began to institutionalize methods of small business and some worker protections.

Thirty years later, as the Great Depression continued, the next era of reforms were enacted; attention to issues of competition for emerging infrastructures prompted the formation of the Civil Aeronautics Board (a precursor to the Federal Aviation Authority), and increased oversight into market entrants drove the relationship between the government and enterprise; "price of entry" reforms limited the business risk for established entities, which helped the government limit the number of business closures that were otherwise threatened by the ongoing depression economic environment. The New Deal brought even more protections for producers, granting them assurance about market access and establishing the terms of competition. Social Security was passed, a monumental social safety net, but the provisions were not extended to agricultural or domestic laborers, a sizable portion of the workforce made up of Black Americans at the time, a mere sixty years post-enslavement. As a result, the law still today excludes classes of workers that were disproportionately more likely to be minority and female.[6]

WHOSE CHILDREN, WHICH PARENTS, WHAT RULES?

It's not by accident that some of the most important reforms excluded Black women or other minoritized groups, or that the arrival of immigrant or minoritized children in the 1900s and again in the 1970s fueled opposition to reforms that were highly beneficial to the majority of (poor, white) Americans; rather, it was a manipulated ploy to harness xenophobic attitudes to limit social support for expansive social programs. Contemporary research has shown that images of Black women in news stories for social welfare programs result in lower voter and public support of such programs.[7] Racism is a powerful tool that can be wielded against unsuspecting policy advocates. At the same time as these machinations were playing out, fewer protections for large telecommunication corporations and loosening broadcasting licensing meant that previously controlled regional telecommunications and broadcasting airwaves would begin seeing fewer controls,[8] and the activities of small-town centers and families would rapidly begin to shift as well. An explosion of electronic stores and hobbyists[9] meant that the era of expanded media content, outside of government oversight, would emerge.

Critiques for regulatory reform increasingly demanded by conservatives in the late 1960s through today often cite the creeping hand of the "nanny state" and the need to prevent erosion of individual freedoms at its behest. The irony of such a metaphor as a nanny state is that in young children the building blocks of nurture that establish a child's capacity for self-regulation have everything to do with responsiveness rather than control. A nanny state would actually be one that provided warm, compassionate nurturance, structure, resources, playfulness, and coaching on how to recognize, articulate, and self-soothe in the face of emotional or environmental stressors. Basic needs like shelter, food, clothing, opportunities for belonging, and pathways to human potential would be readily available to all, rather than aspired to by all but available only to the sons and daughters of scions. It would look a lot like what the Great Society reforms tried to put in place.

What we confront in efforts to regulate TSM today—in our homes, school districts, and society—is the result of the deregulation and push for privatization that unfolded in telecommunications and broadcasting from the 1960s through the 1990s. How the United States structured its management of airwaves—providing licenses in lotteries and requiring those with them to establish audience with increasingly reduced oversight into content and programming—differed greatly from the approach in Great Britain, where hardware providers were granted commercial access, but where airwaves and programming were regulated by national agencies.[10]

Powerful in American myth-making is the notion of individual control, embodied in our concepts of masculinity and "will power," although not purely a masculine conceit or value proposition. Battles to limit PBS subsidies grew in the 1980s, as an increasing number of homes were able to access cable television. Some content would be limited to appearing in late time slots when on broadcast stations, but cable operators would have far more latitude in their programing. A parent could control their child's exposure by either not having cable or not having a television, but it would be a marketing tool to offer more control than that, and when it was offered, it would be a paid add-on for parental controls on remotes and cable boxes, not a standard feature required by regulators.

In this environment, computer science innovations begin meeting telecommunication and broadcast innovations. Computers and sales strategies shifted from corporate sales to consumer sales, and cable and telecommunication lines brought access to them through modems. The ability to market home computers, like the one Seth and I would use in the living rooms of the late 1990s and early 2000s, grew out of changes in telecommunication regulation, allowing cable companies to use existing telecommunications infrastructure, and the increased access by home computers led to increased investments in infrastructure by telecommunication and cable providers, who now also relied on the federal space program to position satellites for increased market coverage.

As parents navigating the tumult that is TSM, it is often invisible the ways in which prior generations of reform have resulted in us holding the bag; how the monopolies on rail and telegraph that Gould built created both the seemingly resilient infrastructure that we continue to rely on for all telecommunications, but also the corporate weight that grants market players dominance over their consumers and channels of influence over their regulators. It's often invisible to us that innovations instrumental in furthering space exploration or valued as national security tools are sources of TSM corporate power, labor designs, and device capacities. What parents receive is not insight into the various large corporate and government interests in NOT regulating the technology our children are exposed to; we receive messages that suggest WE are the keys to the outside world and, therefore, that we alone are responsible for how and whether our children engage.

Furthering these types of individual-responsibility arguments are techno-solution sets offered by companies either commodifying existing technology or innovating as a method of market entry. App timers, the ability to dim the colors of one's smartphone, and apps that threaten to self-immolate if other apps are engaged as a way to incentivize limited phone engagement are all part of such individual regulation approaches enabled by the technology itself. These tendencies toward individual control extend to legal, regulatory approaches, both for parents of children, who are seen as exclusively responsible for children, or who must surrender responsibility of children, and the cultural and social decision-making regarding them. Should they wish to exist as anything other than isolated islands within the ocean of community they otherwise reside within, they must concede control to the computing machinations of the education wave they and their children ride upon.

Horwitz's work emphasizes the ways regulation comes together. Social opinion, industry, organized labor, and politicians in government negotiate on the safety and financial protections that will appease those pushing for regulation without alienating those who need to agree to be regulated. Leaders are motivated to accept and advocate for reform based on the political economy that comes from the public opinion they garner. Protests are powerful signals of political economic shifts, and when combined with industry and labor, they have consistently been effective in granting the political permission that politicians seek, highly sensitive to public opinion in that it transfers to financial support for continued public office. Because of this, press and media are frequent tools in capturing, swaying, and highlighting public sentiments and shifts in attitudes toward regulation. Russia may have been behind the escalation through TSM of America's social discord around race, class, and gender inclusion, but the history of regulation and its failures in America was already playing out along similar themes. Russian networks drew on existing racial infrastructures in establishing their interference just as cable operators drew on existing rail infrastructures in establishing themselves.

In the public square, TSM regulation is often framed as being about individual TSM choices, by selves and by parents. It is logical, of course, to consider a parent's purchase of a device, the provision of internet connection, and the allowance of ungated access and use to be at the center of whether and what types of TSM use are appropriate, or too much, or not of the "right" kind. These are often the debates we allow in our public American square regarding children, partly because they align with our notions of mothering and its all-powerful ability to singularly shape, protect, mold, and refine American children. Never mind this has never been demonstrated. Those with interests in capital growth can easily use individual freedom arguments about violating freedoms, while the most vulnerable and visible on a given TSM device or community will be subjected to the harms of others' freedoms disproportionately.

The NH primary being so pivotal locally gave me opportunities to hear from candidates, and it meant that any industry wishing to have influence at a national or international scale could easily establish a hyperlocal high-touch network of private dinners and privately held accounts with which to arrange one's trip to the Granite State. It also meant that a national headline, like the one that prompted the hyperlocal hate list, are likely to have impact and amplification. What do you do with attention? If you are a seditionist group seeking to create a Free State, for instance, such a headline is both a signal to those seeking such homogeneity, and an opportunity to send signals to those valuing or simply existing in heterogeneity. Though driven by different visions, something about these ingredients seems to make the state fertile ground for those seeking seismic shifts with global repercussions; the monetary policy deals of the 1940s,[11] the machine learning plans of the 1960s,[12] the biomanufacturing of the 2000s[13]—it was all possible in the ballrooms and barns of the granite-filled hills of my state. Regulation and innovation begin with discussions of powerful interests in small states with large white politically engaged populations, like those found in Iowa and New Hampshire.

TIME, TRUST, AND TRADE

It's very easy to just say there's one, you know . . . This is the number of minutes and that's it. Um, that doesn't serve us well, it doesn't serve kids well. Um, and so, how do we help parents and families be who they are, and within that, make conscious decisions around using media well, and

not beat themselves up when they have a weekend where
there's more media use than they had in mind, but that's
how they got through the weekend.

—Dr. Chip Donohue, founding director
of the TEC Center at Erikson Institute and
Senior Fellow of the Fred Rogers Center[14]

My conversation with developmental education researcher Chip Donohue reveals the fine line he has walked as preschool children emerge from
their homes and return to classrooms. What possible meaning can parents make of screentime advice after two years of isolation and increased
technology engagement, and is there any point in directing attention to
such trivialities when so many more important issues concerning a wider
preponderance of families exist? Any parent will tell you, of course,
after such a weekend binge of TSM, disconnection dysregulates, and
dysregulation disconnects us from others and our own experience. TSM
connects, but it can also disconnect, and it dysregulates through discon-
necting individual users from their reality, community, and autonomy.

I lived in New York City in the late 1990s and early 2000s, between the bubble years. In those days TSM was moving from candy bar
cellular phones to flip phones with texting. There was a print world,
glossy, and large pages on every green newsstand dotted along the sub-
way entrances across the city. Media was still highly NYC-centric; even
Hollywood and cable television productions in California or Toronto
would feature and amplify NYC stories. In the late 1990s, the NYC
media was abuzz about a film that had come out, amplifying the effects
of climate change and predicting peril. Technology connection provides
witness, a shared reality, which leads to shared narrative. The evils of
TSM as it stands are not about this connection or witness; they are about
this tension that comes from opacity to the narrative and reality. Today,
my state and others are working to limit access to reality by limiting the
access to education possible and history available in classrooms, from
preschool through college. They are doing it in the guise of protecting
all students from divisive concepts, but their encouragement of boun-
ties and hotlines makes it clear which education and which educators
are likely to be challenged. It's not a mistake that TSM-driven social
division is easily leveraged in the effort to limit educational equity for
minoritized children the year after international protests in support of
Black Lives Matters. It is the design racialized political systems have

been laid upon, and TSM allows familiar scripts to be repurposed with alarming efficiency.

Attachment regulates, but it is also the case that disrupted attachment creates barriers to regulation. Detachment disrupts coherence, connection to meaning and expression that is part of one's history and self, and requires new attachment to form while the ruptures of failed attachment are unaddressed. In humans, this looks like parental separation; children can be reconnected to caregiving adults, but the disruption still will have long-lasting effects on the child's sense of self, pattern of relationships, and expectations for their future as I experienced as a child with frequent, forced family separation due to racism, sexism, poverty, and violence. In TSM regulation, dealing with policies or practices without attention to attachments, this looks like individual companies detaching users but allowing networks that amplify detached user messages to continue to flourish or be memorialized by the detached group. Coherent apperception of the communications then shared becomes disconnected from the community to which it was attached. What does that actually mean and look like?

When seditionist Trump supporters were pushed off Facebook in November 2020 and onto other newly formed-for-sedition platforms such as Parler, the remaining community within Facebook could hardly understand the nature of the posts they were witnessing being reshared on Facebook from Parler (as screenshots and reposts). There was no coherent way for the non-sedition Facebook user in isolation to understand or appreciate the meaning of the messages that, despite deplatforming, were still highly visible within several other related groups. In this series of actions on the part of Facebook, they monitored and filtered content that might be seditious or inflammatory, and they chose "engagement over safety" at the individual user level.

Beyond individuals, we often seek regulation from institutions, whether that be the FCC or SEC, both of whom tend to have purview over TSM in the United States by virtue of their interest in communication technologies and in their interest of financial regulation. Institutional regulation, however, is hampered by itself; if the FCC regulations challenge the value held in a company, then SEC-related pressures can shift the intention of the regulation. If pressures on transparency and investor protections require SEC regulation, there will be an impact on the availability and accessibility of the TSM being regulated.

The machinations of the telecom breakup of the 1980s and the cable wars of the era are easy to spot as instrumental in yielding the TSM today. What is harder to appreciate is how more recent regulation shifts are yielding TSM that will further remove control and regulation from families and individual users, or ways in which our already connected social and financial infrastructures have become international targets of global actors seeking national wealth for their small countries. The financial industry is highly regulated; capital markets control capital, and that is a job for the currency creator, the United States government. Technology innovations, and economic shifts, have made such centralized control fragile.

In the years after the Great Recession, technology company investments increased as investors sought less-regulated and high-return investment over bonds that were continuously depressed in order to uphold overall markets. By 2015, high tech and financial industries began targeted investments into "fintech,"[15] an amorphous category that covered everything from peer-to-peer payment systems as well as alternate currency markets (crypto). Over the one-year period between 2014 and 2015, investments within Fintech increased by twofold, and the increase now covers more than a tenfold increase in investments.[16] In 2021, the evolution of this growth resulted in more than $29 billion in investments in non-fungible tokens (NFTs), which represent both the next wave of creative licensing and the largest money-laundering opportunity for international interests. Regulation of such is apparently left to individual artists, who have limited ability to retrieve poached digital images now tied to NFTs they did not authorize (or profit from), and individual buyers, who have no recourse in correcting financial calculation errors as they attempt to calculate obscure valuation exchanges. Buyer beware has implications for industries and investments in ways little appreciated by average observers of the phenomenon ushered in by the merging of traditional finance and the development and deployment of blockchain technologies.

IN LOCO PARENTIS

Despite this flow of wealth and the national vulnerability of the connections between our infrastructures, the pandemic revealed how few low-income families were actually able to reliably connect to the internet in order to support the remote learning of their children. Attention to apps and the controls parents might rely on gave way quickly to recognition that the digital divide was indeed wide. And it was beyond simply device or internet access.

> One, it [the pandemic] revealed how many families were under connected, or not connected at all, an, an unbelievable inequity around access to digital, um, tools. And even if you had access, did you know how to use them? And, and do you know how to help your child use them? And do you know how to support your child's learning at school?[17]
>
> —Chip Donohue

For parents, institutional regulation of TSM comes in the form of "loco parentis," which extends the rights of parents to choose and control their children's environment and shifts to the non-parental adult an ability to act "in place of" parental authority. This was a necessary convention in the days of the Progressive Era reforms, which had to simultaneously reconcile the right of an individual family to control their child with the reality of compulsory education where non-family adults would have to retain order and control. The parent would be responsible, to the point of incarceration, for their child's arrival and retention in the school building, and the teachers would be responsible for the child during that period away from parents. This was also required in institutions such as juvenile reform facilities, where parental authority would be all but terminated and the "state" would take the child as a ward, granting them authority for control. When TSM enters, it's often first because we presume some achievement possible through it, then we expect access for more so that the general community can achieve, especially if we perceive their need to achieve as being tied to our own economic wellness.

The tensions of these realities played out on a small scale in my home when one of my oldest children began attending a day school many years before the pandemic. Our eldest wanted to attend the small school, a dramatic technology shift from our public neighborhood elementary school, which still relied on printed worksheets and the occasional composition book for student work into a middle school that had adopted iPads for their students to appeal to wealthy parents expecting a high-tech educational experience, and for a school networked with other private day schools and therefore wanting to keep pace with innovative curricular tools rapidly adopted by their youthful and rotating staff. For us, however, it was a radical break from our homelife.

Although we had adopted iPhones for ourselves, and eventually each received iPads related to work or board service, we had not until that point acquired any technology, never mind some like the iPad, for the individual children in our household. By virtue of sheer numbers (of children versus of devices), we had forced regulations on screentime, app use, and games played. With the introduction of a personal and portable device for an individual child, the stakes shifted. Atop these shifting stakes was the reality that concentration, task management, and other metacognitive executive-functioning skills were being developed at the same time social skills and socializing online was being explored. The combination resulted in many nights of arguments and sleepy mornings, as the tension between enough time to work on challenging homework and too much time that allows non-homework activities devolved, so did the grip we had on what tech, when, and for what purpose. Everything was available, everything was necessary, and nothing we wanted for our child mattered in the demands of the educational environment we had elected for her.

A comment during my interview with *Internet and Society* researcher and journalist Emily Robin Dreyfuss distills the experience we were left with in our impotence:

> Consumers don't have a lobby the same way that the companies do. Like, we literally are in such a vicious cycle, and, like, a closed system where what people do is they go on Facebook to talk about how they think Facebook should be regulated.[18]

Other historians and scholars offer credence to the domination and immutability of TSM influence and power in individual family life, but increasingly in the public square, not only in China, where uses of face-recognition technology and social media analytics are combined to create mass surveillance policies, but also within the United States and wherever TSM is deployed. The 2020 documentary *Coded Bias* by Shalini Kantayya follows MIT Media Lab artist fellow Joy Buolamwini in her quest to create a mirror that will greet her with daily inspiration, but which won't function using the existing technology because the sensors won't detect her black-skinned visage. What begins for the viewer as a fanciful art project with technological challenges becomes an indictment of policing, computing, and the commercialized and militarized deployment of flawed technology that disproportionately harms those least able to know about its deployment.

The social and legal implications of a world that is increasingly reliant and incentivized to run off artificial intelligence built into sensors, social media algorithms, and even employment review systems are significant. At the same time, the human knowledge and skill needed to manually manage[19] the amount of information available to computing systems is beyond measure, and here is the central conflict of regulation within TSM deployment; increasingly to create usable tools responsibility for content moderation, predictions about where help is needed in community policing, risk and restraint management in criminal justice cases, clinical judgment in healthcare delivery, and even assessments of child development are completed by coded actions opaque and inaccessible yet that will drive localized actions under the guise of care surveillance enabled by an elusive "Internet of Things." We the public, however, are assured that technology sector growth is an economic good, one that will begin to return the high-paying, middle-class jobs that once supported men and their families. Even as Emily Robin Dreyfuss and I discuss the implications of the tautological options for redress apparent to most TSM users, she delivers the argument that sticks

> . . . and, like, tech regulation, I just think it's all tied up in, like, larger social, um, regulations. Like, we, we wouldn't have the loneliness epidemic that we had if we had a better education system that set, you know, young men up, in particular, for jobs that would pay them a living wage. Like, all of these things, I think, go hand in hand.[20]

In discussions of regulations around TSM when parents have limited control over its introduction in the educational environments, and when the public has little control over its deployment within the community, and technology workers themselves are granted little control over the use and amplification of their work product, it seems there is no simple answer to the opening question of this chapter. "Regulation, but whose?" indeed is the lingering question individual adopters of TSM are navigating, but often they are not doing this with the knowledge and information that they in using the various TSM available are themselves the beta-testing, money-generating, social spreader they imagine large technology companies are.

Pandemic has finally pushed us toward the final frontier of *parenesis parentis*, when regulation is granted not to parents, or in place of them, but as part of the responsibility of care that government demands to preserve its people's well-being. We saw this type of regulation unfold in the late 1990s, as states began bringing suits against tobacco companies for the harms their products and advertising techniques were bringing. Zooming through interviews in the fall of 2021, I am considering what regulation can look like just as congressional testimony from Facebook and other social media leaders begins anew. This time, once again, there is scrutiny on Facebook, which will by the time the congressional meetings are wrapped announce it has rebranded as Metaverse, happening a year after releasing "Meta," a platform for VR avatar-based meetings and community. In every interview I conduct, interviews with leading experts on the developmental impacts and the potential of technology used with and by children, youth, and adults, we are all struggling with identifying our own sense of balance, regulation, and purpose in our necessary reliance on technology to participate in society in ways matching our preparation and skills. It's obvious there is no way forward without the tech, and all it has wrought and will yet do. And it's also obvious that even with our collective expertise, insights, and convictions about the centrality of human, ecological, and social dignity and the interconnectedness of this that exists none of us individually even in our posts feel much positioned to influence the frameworks that will be used to discern the tech that is still coming, and the human changes it is creating.

Beyond films like *Coded Bias* are the headlines of tech behemoths mistreating the very accountability talent they recruited and empowered, disempowering them just as their voices gained clarity in what fixes and reversals are needed to preserve human rights, the dignity of work, and the accountability mechanisms of corporations. The intensification seems connected in some ways to other headlines trending; ones about the skyrocketing values of various cyber currencies, about towns shaken down for millions in ransomware attacks, and about misuses of personal information on company manufactured, if not issued, devices. Within a week of the twentieth anniversary of 9/11, the UN issued a report[21] calling for the halting of AI programs by governments and entities unless and until human rights can be preserved when deployed. This urgent demand, complete with data supporting the increasing and accelerating risk, is piling up. What began as a collection of worried self-observations within my own nursery and case vignettes from my lab is tracking on many levels with the increasingly documented concerns of TSM, tech company monopolies, wealth inequality, and the concentration of resources and opacity of TSM empowered or deployed AI.

When it comes to regulating the algorithms that increasingly provide healthcare, education, food, water, housing, transportation, and legal services, we must take seriously the threats and opportunities before us in slowing development pace and critically examining adoption; at least there is global clarity in the call for care with AI. A call for AI surprised me in a different way when, in late 2021, I was invited to review projects proposed to support the increased adoption of machine learning in order to promote mental health treatment innovations, or otherwise support the biomedical workforce professionalization. I don't create algorithms; I am not a computer scientist. Something about the STEM education projects I have recently been submitting for funding at various levels has likely drawn some attention to my cultural emphasis in bioethics and STEM education. Still, I wonder if accepting the invitation to review should be met with a greater hesitation than I experience. Should I ethically examine AI projects, if the presumption is that AI would be used and retain availability and structure? Does ethics require deconstruction rather than revision?

Similar questions arise when invitations to my lab to help create content for an employer AI-driven platforms wishing to use custom diversity, equity, inclusion, and belonging (DEIB) modules to boost health equity for marginalized employees. Would such precise information delivery actually be used as a Trojan horse to extract self-selection behaviors as a way of constructing risk profiles for the employer to use in preparing their insurance renewals? Can creating the stickiness a platform needs to extract data from users contribute to the threats to access and equity that TSM enables and has empowered? I had thought that navigating the lab work between the conference in California where I meet Lyon to starting conferences of my own examining technology and community intersections would make me more clear-voiced in the needed regulations of TSM; but each innovation cycle introduces some new, but long predicted, incremental scraping of information that is then often misapplied based on erroneous data assumptions.

How do we begin to embolden regulatory opportunities when the scope is so often directed by our financial technology, health, and astronomy innovation funds? We feel at a loss to stop the roll of what's already rolling. Optimism is present, insights from seeing resilient innovation and public focused education around the power and potential during pandemic remote instruction and work are starting to inform practice and policy in businesses and districts. Even so, the technology touchpoint represented at a time climate change is tied to fuel investments, when global currency markets are linked to unregulated and unprotected cryptocurrencies (themselves linked to digital identities controllable and interruptible beyond scales we had considered—except in our science fiction) is indeed powerful when combined with the forever traceable genomic and digital user prints that we are connected to, and with the increasingly affirmed and observed inequities in algorithmic and ecological justice access already documented and ever widening. We must recognize that ultimately TSM impacts nations by and to a far greater degree than any given YouTube video seen unexpectedly in any given family home.

When the attachment system itself is being damaged, whistle-blowers emerge, spiritual advisors sound alarms, and the ecological frameworks around a child that Bronfenbrenner so astutely articulated begin to shift their focus from the child to the system. We have arrived with an awareness of the interconnectedness of these telecommunication, governmental regulatory, and human regulatory systems in ways we have not been able to appreciate before. This is the moment when TSM regulation will be realized; but just like reforms of prior ages, it will not be shaped by the needs of children but by the demands of corporations and to the limits of the political appetite for corporate control and government regulation, and tied to the limits of inclusion that white majority Americans will tolerate. In other words, we can expect this era of reform, already taking shape away from administrative controls, to fall short of the attachment imperative of that calls for responsive, attuned technology regulation stemming from an ethics of care[22] that preserves labor, human relationship, and individual privacy unless these are tied to national security or corporate interests. Or, unless great political protests can persuade those elected to demand more.

• *14* •

Socializing Social Media

*A*eronautics, electronics, and cable information transmission transformed war from the start through the end of the twentieth century. By the time of the *Challenger* shuttle explosion in 1986, which left such an impression on my classmates and me, these same technologies and their infrastructures would transform peace, ushering in the end of the USSR, establishing complex international academic partnerships, and laying the foundation upon which the nucleus of our current and emerging (Web 3) TSM would be established (and battled over). A year after the *Challenger* tragedy, a set of duck statues brought joy and healing to a city and set the unexpected stage for the peace between world powers.

It must have been some comfort for the mothers of the 1940s, reaching to read a story to their preschoolers, and finding within its pages a story of competent mothering in the face of family separation, community in the face of urban isolation, and security in the face of global deterioration. Such were the themes evident to me as I reread Robert Mc-Closkey's *Make Way for Ducklings* some eighty years after its 1942 release. The sepia-toned line-art story depicted a pair of Boston city mallard ducks who must scout for the perfect location to raise their many ducklings. The ducks settle on the banks of the Charles River, only to foray across busy streets into the Boston Public Garden and back again. Once the ducklings (Jack, Kack, Lack, Mack, Nack, Oack, Pack, and Quack) are swimming strongly, Father must go away (for reasons unknown) and promises to reunite on the small island within the man-made duck pond where the city swan boats ferry tourists in leisurely circles.

The illustrations of the urban landscape from a flying duck's–eye view, combined with depictions of urban helpers like the attentive

179

policeman, are of a different time and tone, one that elevated modernity, human possibility, and the orderly humane commune possible between humans in their great urban ambitions and nature in its persistent resurgence. For World War II–era mothers, the metamessage, that with care, order, and confidence they could manage many little ones while fathers were away, was likely not lost on their reading of the work. The book wasn't designed for any of these higher aims; indeed, McCloskey would say he just drew what he knew and remembered from his art school days walking across the Common on his way to class. Still, the story held special sway during a lull in white feminist organizing between the suffrage feminist activities in the 1920s and the workforce feminism that would come in the postwar years. For children in the late 1980s, a city planner with children's use of public space in mind and a sculptor would work to help bring the story's characters to life as bronze statues on cobblestones nestled among the curving paths of the Boston Common—home again.

Not far from a matching pair of those very same statues and cobblestones, on the grounds of a former convent, rest the remains[1] of Lev Vygotsky, the Russian psychologist who helped contextualize how it is that children, in relation to adults and peers, learn and develop their conceptual understanding of their social world. The Moscow ducks were gifted in 1990 by First Lady Barbara Bush to Mrs. Raisa Gorbachev as a way of fostering goodwill between the two nations as they finalized the INF treaty—setting the stage for nuclear nonproliferation and marking the end of the Soviet Union. At the 1991 dedication ceremony, the wife of Prime Minister Gorbachev beamed, and guests in attendance of the statue's unveiling heard of the promise of wealth, prosperity, and cooperation that was soon to be available to all those in attendance as a result of their government's willingness to put nuclear arms races to the side in favor of domestic development.[2]

Ideas are powerful, images that communicate ideas are even more powerful, and ideas, once loosed, have a funny way of shifting beyond their start and ending in places least expected. It may have seemed somewhat confusing to consider what small bronze duck statues had to do with promoting prosperity, and it certainly must have baffled how these two conversations had anything whatsoever to do with ending the threat of nuclear war on earth, and preparing for war with alien life forms instead,[3] yet the moment would not have been possible without these factors being ideas firmly in the minds of the men who prompted their emergence. The duck statues were simply a material representation of the domesticity, natural world, and maternal warmth that awaited anxious readers of the 1940s and rival superpowers of the 1980s.

Vygotsky appreciated the power of stories, and he wrote much about language, cognition, and ways in which adults, peers, and children's creative interests combined to create their own unique learning contexts. In these individual contexts, children bring forward their curiosity and knowledge, while the zone of proximal development (ZPD) between their understanding and ability is bridged through adult and peer interaction. As children are attuned to by teachers, support for the skills needed to bridge the individual child from their limits of understanding to the edges of performance becomes possible with adult and peer support. It is through this exchange and the extension of the child's curiosity with the adult's ability that a child learns. Knowledge, argued Vygotsky, is constructed through dialogue, both between a child and their educator or parent, and between the child and the knowledge, for once captured, the child will then need to assimilate the understanding with their prior knowledge and develop a new conception of their world or themselves.

During play, Vygotsky argued, "children are a head taller," meaning they are interested and motivated to explore and engage at the level above their actual developmental point. Play is an opportunity to learn by extending the ZPD without having to truly perform within that zone.[4] By practicing and rehearsing for the developmental level ahead of them, children gain confidence to extend their curiosity ever more, and in doing so then become competent for the stage of exploration available to them as they condense and confront their knowledge.

Media requires mediation, and that is the process of socialization that assists the child in their meaning-making. With families, then, we can outline the zones of proximal development that shape TSM for a given family or child. There will be the parent's ZPD as it relates to both the TSM being considered, but also their ZPD for their child, its contexts, and the context of the TSM content they are seeking to mediate. Mediation without attachment is programming; inputs with intended impact but that are not interpreted through relationship will not translate into understanding for a child, even if they are mimicked and repeated. Mediation through relationship does not have to be in person, of course—my generation mediated the meaning of myspace and

Tumblr by engaging in online relationships, organizing into interests and idea based groups, and through these affinities found and created the mediums. Even so, our online world creation was anchored to our offline relationships since these would initially filter our views and attention. It is the same now as my children connect with their relational groups through TikTok and Snapchat. Culture contextualizes, and so attachment provides both the ground and its limits as we consider ways we collectively might shape and socialize the TSM in which we and our children engage. The dramas within TSM landscapes, however, are often conflicts between contexts, mediations between meanings that are not universal but specific. "Story," author Lisa Crockett writes, "is the language of experience."⁵ TSM brings us many other people's experiences, offers ways to reshare our own, and can also offer limited connection to others' stories without specific intention, most especially the stories of those least empowered within a given space. As we are learning more clearly than ever before, stories are not singular.

Since stories are not singular, we have no reason to presume families would be. How they mediate, what approaches they select (when selection is possible) are universal only to the degree that selections are mandated, either by regulation or by availability and design. A funny thing about parent controls is how few companies thought to adopt them before troubles brewed for the industry, how proud TSM execs are to announce them now, and how little parents often understand about deploying them once they are available. Understandably, families are often desirous of industry mediating the information between their children and the content that the TSM delivers; here we see the most attempts—and the greatest satisfaction—with mediocre results. It's one thing to work on your own child's regulatory context, another to consider the internal organization of a given company or agency, and a whole other to contend with the industry interests that have little care outside their continued ability to operate at profit without having to invest too much capital in constraining their TSM tools. Such battles between industry innovation and parental expectations are not without attention, and while this book may outline the various political, social, capital, and developmental forces in the way of effective TSM improvement, there are individuals who work actively to make TSM for children useful, contextualized, and developmentally enriching. There are limits to how well represented the needs and expectations of parents in various social contexts may be in these efforts, but they do exist and have yielded positive outcomes when it comes to shaping educational applications for TSM.

MAKERS, SPACE, AND KINGMAKERS

Recognizing that games and tablets had a potential to change others, a group of app developers began gathering annually. Initially a handful, the event grew, and today anyone wishing to attend can register and self-fund a weeklong exploration into emerging educational apps informed by developmental researchers and educators. The conference, Dust or Magic, was brought up by both Chip Donohue and Lisa Guernsey (originator of the Context, Child, Content guidance) when I spoke with each of them over several months in late 2021. Chip reflects on the early years, sharing the monocultural perspectives apparent in the app designers, mostly young men, in their early to late twenties, single, white, and educated (by self or by Stanford, the early bifurcation of the tech workforce)—guessing at the simple game mechanics of their youth and embedding them into "educational" technology. To hear Chip Donohue describe the event in its earliest years is to hear Emily Robin Dreyfuss describe her early tech journalism career, experts with an interest in helping the technologists shape the future for good. Within technology journalism, in Emily's case, that involved taking complex biomedical, computer science, and telecommunication commercialization in the form of the technology industry and helping non-computer-science readers understand the big deal that was the emerging technology of the early 2000s. In Chip's recollection, the beauty of Dust or Magic was not to critique or repair misguided apps, but to support, elucidate, and grow a developer understanding and appreciation of technology for young children that took into consideration their Theory of Mind and Zones of Proximal Development, and to cooperatively shape the educational contexts of their bioecological systems. Each of them (Chip, Lisa, and Emily) highlights the roles of developmental specialists, educators, and journalists to enhance and encourage promising approaches to engaging young users with quality technology.

The collaboration between child educators, developmental re-searchers, and app designers that was facilitated by events like Dust or

Magic was part of an iterative design process familiar within computer science and engineering. For Emily, her task to translate and communicate technology's promise and perils was to focus her reporting not only on the apps themselves that might come from a given developer, but also to expose how industry investments were not spread evenly across innovator categories, and offer critique regarding which development team was most likely to have the social and economic capital to attend such an event with industry contacts, and maybe amplify readily available partners for deepened work.

Of course, a venue like Dust or Magic, while not intending exclusion, has barriers to entry and communities who won't benefit in their app development process from expert developmental research. And this matters, not exclusively to the specific Dust or Magic conference community perhaps, but to the children and families who will use apps developed there or elsewhere. The absence of developers with certain types of social and global experiences, the limitations for educators of specific learning communities, and researchers with minoritized identities or research expertise impede the degree to which emerging technology can be informed for communities. As the emerging technology is laid upon the infrastructures and ingroups of both the early years of TSM and along the early communities of influence, we see biases and blockages that exclude even without intention, but that when coupled with exclusionary or exploitive intentions can be devastatingly effective. Algorithms informed by use practices, data scraping, and aggregations will—by design or default—replicate such practices. Serena Dokuaa Oduro and Andrew Strait warn of this in their essays in the book *Fake AI*, writing with clarity that this is not a small thing to recognize. When planning and iterating powerful technology like that found within education apps, it is imperative that attention is clearly paid to those not present:

> If any historically marginalized group is not mentioned that group will be harmed. Any marginalized group can ask these (and more) questions and it's important that we do. Current analyses of fairness in AI often flatten the nuance between—and variety in—Black women's experiences. While flattening identities may make it easier for AI, it harms Black women and leads to further AI disillusionment. If the field of AI is to gain public trust, it will have to prove its benefits to all people.[6]

When not all people are, or can be, in the room where technology is created and deployed, this is a large task indeed to conceive.

It's understandable then that our TSM controls are frequently focused on the individual: device, user, home, or system. We see the obvious solution (just say no) as the only solution, however, too often, and we do so at our own collective detriment. But the attachment between a parent and child, one that could be strong and still go wrong as my own mother's did during those months of grounding, is one that functions as a coregulatory process. Any individual regulation observed is coregulatory in impact, and likewise our thinking regarding technology and regulation needs to be as connected, nuanced, and responsive as the attachment system that has brought a given family and humanity to this point.

When we consider additional approaches to regulation, we can identify similar patterns of individual responsibility without rights, institutional responsibility without obligations, and industry responsibility without constraints play out. Moderation is the answer available, both for users who are encouraged to self-limit their engagement and exposure to platforms and activities; and for companies, who can outsource human judgments but to do so at scale and with the cost margins their investors require must do so to humans who can earn less than the prevailing wage where their headquarters are set. While human moderation is more comforting for users, context is not universal, and understanding the unique bioecological contexts of the user posting and the content moderators making judgments cannot be relied upon. We can't lose sight that moderation is a human task, done by and for humans. In his *Fake AI* essay, former Google content moderator Andrew Straight reminds us:

> Content moderation is difficult work, often exciting but occasionally damaging to the soul. The platforms I moderated were (and still are) used by billions of people across the world as libraries, news sources, and public squares, but they were not built for these purposes. Rather, these platforms were designed to incentivize user engagement, content sharing, and mass data collection.[7]

He continues, urging more awareness of the implications of human moderation, when as Vygotsky's work and Urie Bronfenbrenner's extension of it should already make this plain to users, yet often it is outside the conception of platform creators who suggest that artificial intelligence can both protect a human workforce and protect a user.

> As with many other AI technologies, the hype around automated moderation tools reflects a misplaced belief that complex sociopolitical problems can be adequately resolved through technical solutions. AI is tremendously helpful at addressing well-defined and concrete tasks, but determining what speech is acceptable for billions of people is anything but a well-defined challenge. It is understandable that we wish it were so—how nice would it be to simply let a machine police our speech for us without changing the scale, practices, or affordances of Facebook, Twitter or YouTube?[8]

We see examples of this failure in global conflicts where the power to share content with a broader community can bring attention to harms by state violence unless silenced by miscalculating algorithms[9]—not due to a corporate lack of care, we are to believe, but from an inelegantly applied AI—and human moderators without a contextualized understanding of the power dynamics underpinning local actions. While examples can be found of failed protections for users or moderation of important evidence of human rights violations, the real progress and growth in the tools parents might consider essential in mediating the human user–driven TSM ecosystem are the tools that threaten democratic freedoms and public safety for many low-status communities here in the United States and abroad. Content moderation creates channels of policing that are outside of the judicial systems established in a democracy[10] and databases of information that can be easily repackaged and repurposed to further develop technology that erodes user freedoms in the offline spaces where users must physically exist.

Of course, families want to limit the risk of harm for their children as they traverse the informational networks available through TSM, but the impact of identity and misidentification are not evenly borne by all families.[11] As image recognition technology further evolves, in part because of the relational and informational connections established by the regulations that have been enacted within companies or as a result of regulations implemented, this deepens. And the storage, recursive revisiting of previously analyzed images itself contributes to ongoing growth of internet infrastructure.[12]

This isn't to say that efforts are not serious or significant; stories of newly programmed TSM that better addresses accents, or accessibility, or complexion do exist, and platform changes have also been executed and supported as a result of the learning that is possible when institutions are confronted with their bias and ecological misalignment. But this is due to unlearning the position of dominance that is otherwise within the zone of understanding developed until the outrage or demand is made. Responsive systems aligned around the strength of attachment respond, but socialization is required, and that can only occur where and to the extent of the mediation that is available between the TSM and the communities designed for—and not designed for—in mind and is no guarantee that good coordination will be maintained as different parts of the overall system progress at different rates and in sometimes conflicting directions.

While space is the imagined "final frontier," earth is the home to which we are inevitably attached, no matter how efficient our rocket propulsion might become. The story of outer space, of infinite energy, and of expanding connection to others have all been fuel for TSM's expansion. Every iterative innovation was a sentence in the story of forces that could extract, exploit, and expand empire used to convince our adoption, rally support, or mute our opposition. Rather than doing this directly and on the merits of the TSM or resource need, however, this was accomplished by reifying stories that supported the notion of independent individuals on personal and society journeys to realizing their, and its, full potential. No stories of interdependence, no realistic appraisal of resource availability and renewability, and no consideration for the permanent attachment and dependence even our most ambitious technological advance would not break for us: We are human organisms in need of the conditions of earth to flourish, reproduce, and share culture. No amount of technology, TSM or otherwise, can erase this reality about the limits and lamentations of our humanness.

Before I learned of Dust or Magic, I waded into the waters of socializing technology as an extension of human interconnectedness with place, in hopes of supporting addiction recovery at a local level. I worked with an app developer to take their existing commercial app based on smartphone-delivered customer rewards through iBeacons placed in local community businesses. Together we explored how to leverage the commercial technology to reward recovering substance abuse patients for engaging in substance-free, prosocial community events (art museum visits, faith community services, block parties). Since we lived in a state with the highest rate of opioid addiction, it was a logical, if not ethically[13] neutral, endeavor.

We never launched the project, but the vision of using positive reinforcement delivered through smartphones via iBeacons in the community to increase positive recovery behaviors was firmly grounded in treatment literature that demonstrated consistently that social engagement improved recovery and functional behaviors. Our technology was an Internet of Things (IoT)–based network that would have captured and incentivized the user who had the app and opted in to the recovery support features we explored, adding into the commercial beverage and appetizer incentives most of the hospitality industry that had adopted the program had desired.

Within a year of that grant, thousands of inexpensive tablets sold between 2011 and 2015 using the same IoT technology to wirelessly connect to the internet through Bluetooth signal were weaponized when one of the worst globally impactful DDOS attacks shut down a local internet company, Dyn, on its verge of being acquired by Oracle. The attack unfolded on a cool October day, and before it ended that fall of 2016, nearly 10 percent of their customers were offline for more than an entire business day. The attack was fodder for the technology press and spurred innovation in the founders, who within a few years would start a new company based on the insights of how IOT-connected devices could be managed within home and business settings.

SUSTAINABLE SOLIDARITY
AS TECHNOLOGY SOCIALIZATION

What was once established to help industrialize information exchange has become its own industry, with each new edge of the industry justifying its own further evolution, creating an invisible but tangibly consumptive engine; consuming resources, individuals, and communities in ways that "cloud" as a term makes invisible. If as contemporary writer Nathan Ensmenger argues "the cloud is a factory," then we can expect feral ecologies to be found in the spaces within and surrounding it.

Donna Haraway, who penned the 1985 feminist technology opus *The Cyborg Manifesto*, and anthropologist Anna Tsing, who wrote the

2021 *The Mushroom at the End of the World*, have developed a way of understanding the ecological impact of humans' contextual experience in the last century and a half since industrialization. While this era might be referred to as the Anthropocene, as briefly discussed in Chapter 12, they find an even greater value in understanding the specific changes associated with the industrialization of labor, one that began not with the mechanized textile cum tech incubators that I live near in my small northern US city, but that began with the sugar and forced enslavement of my Carribbean ancestors: Their name for this is the *plantationcene*, which helps to identify the changed ecology of simplified environments maximized for specific labor and production. If the internet is a cloud, the cloud a factory, and computing ubiquitous as the emergence of Web 3 portends, then the plantationcene created for TSM is within us.

Socializing TSM then will mean decisions about production, extraction, and mechanization that seem far removed from the daily waking and living cycle of a given individual or family. But these realities are hardly new. While Haraway and Tsing's concept is developed in the post-2010 context, the observation of permanent interconnectedness between human and computing was a refrain in Donna Haraway's interview with *Wired* writer Hari Kunzru published in February 1997 and reveals the emerging understanding of the connection and interconnections between military research, research inspired by the prospect of human space exploration, and advances in human/machine interfaces. Long before social media was keeping parents up at night, Haraway was quoted in the article as saying, "Technology is not neutral. We're inside of what we make, and it's inside of us. We're living in a world of connections—and it matters which ones get made and unmade."[14]

Connection is a theme in Harway's work. In molecular biology, contact zones are areas where one molecule and another molecule come into contact, with one overpowering the other and creating something new, or destroying something old. In terms of the plantationcene, the contact zone can be understood as the place where communities and individuals of differing power interact, with implications for freedoms, rights, and options for behaviors that are contextualized in the power relations between the groups and people or systems involved. To ground this further, we must understand ourselves and our families as operating within the plantationcene of TSM, and we appreciate that our engagement with TSM as a user or parent is embedded in power differences; between ourselves as individuals and corporations protected and permitted by laws of the physical location the corporation is choosing as its base; we like the children who benefited from Head Start are operating within systems older and more established than our individual sense of rights otherwise might allow us to imagine. Like products from plantation production, and like humans improved on earth for space travel, we are dependent and complicit in our generation of product and consumption within these contact zone arrangements.

It's common, of course, to consider individual users as powerful; Elon Musk in his quest for space[15] sends economic markets spiraling with a single tweet, and we see the effects of the January 6th insurrection and the actions of companies to remove users, like President Trump,[16] for inciting the uprising with their posts. The technical process that had to unfold for a post or a tweet in January 2021 is remarkably similar to that which would have been required to post in the earliest days of internet life; a modem connects a phone signal to the internet, a router connects the internet connection from the modem to your device, and a bridge allows signal priority for traffic and from a device, allowing it to bypass other signals awaiting transmission. Local nodes can spread information between users; this was the insight and reality of the earliest decentralized networks of users that allowed even small countries access to internet before the commercial opportunities for users in lower-income countries had national options for connection.[17] It is, however, the central nodes of connection that scale signal, ideas, and practices. TSM allows scale through its controls, spread through local nodes treating them as scale quality input. It doesn't matter if there is not quality in the transmission; the network functions to scale once central nodes are activated. Emily Robin Dreyfuss, in our interview reflecting on the changes within TSM and the technology media coverage, describes the translation of this into online experiences, saying:

> I mean, internet and social media is a place for people to express their extreme unhappiness, and their sorrow and their rage, and then social media has created a speed and scale that allows that rage and sorrow to spread so much faster than old systems like the press, like the healthcare system— . . . like the government can['t] possibly keep up with [that speed and scale].[18]

While there is often a recognition among technologists that innovation speeds in a given context exponentially, as opposed to geometrically, this same understanding does not translate into education or governance formulas. Staid models of proposal development, community approval, or curricular design are configured as the deliberative ground upon which technologies generated are to be considered for adoption. When I sit in on a curriculum committee in the early 2010s, where some departments in my small liberal arts campus are clamoring for automatic grading resources, ones that capture student writing content and then auto-compare to other student writing samples in an effort to sniff out plagiarizing students, it's clear the speed of innovation and marketization has outpaced the governance structures available. One academic department that perceives its workload and mission as imperative to continue over enrollment was sufficient to overturn considerations of student's intellectual property rights, the realities of renumeration, and the protection of students from the plantation model of corporate extraction. No matter that the college was prized for its commitment to civic participation, democracy, and the humanities—the simple mechanics of "students need to write more papers, but professors wish not to grade more papers" was sufficient justification for accepting the anti-humanist and antidemocratic user policies, and to even pay for the privilege through monthly subscriptions. It's still in use today, with no mind to inculcate new faculty on its judicious use (if there is one) and no mind to cultivate in faculty alternate measures of writing and performance. That would require investments in time and resources that are harder to propose and govern than a simple company agreement contract.

Is there a way forward then, when the technologists, in concert with development experts or not, can create the next glittering thing into which ourselves and our children will be drawn? Would education make a difference given the experience on my campus in microcosm that must be increasingly true of other settings? Some think so, and efforts at an emerging, intentionally interdisciplinary approach to cultivating both deep content knowledge expertise while preparing for the realities of the individual roles that ultimately govern the launch and rollout of such innovation are emerging. If the effort to expand the approach is successful, it will mean moving from the "critical thinking" and problem-solving emphasis that the entrepreneurial-driven ethos of the late 1990s and early 2000s brought to college campuses. It will also mean something even deeper must occur. The lofty and inaccessible classroom spaces of management training or college degrees won't be where the revolution to socialize our technology will rest. Indeed, as Vygotsky's 100-plus-year-old theory itself would attest, our ability to learn and grow is directly tied to the knowledge and growth edges of those supporting us through our ZPD. We can only know what there is to know within the relationship where learning occurs. To go beyond, we must be able to learn in relationships that we are not in contact with, or, rather, to transgress the contact zone of our ZPD in relation to those whose power and privilege we are dynamically engaged but often invisible to us.[19] It's an insight Columbia SAFE lab director Desmond Patton shares with me in his interview when I ask if our mutual interest of inclusive, nonexploitive TSM is feasible, if maybe the horses out of the proverbial barn have gotten just too far. He doesn't think so. "Is it really possible?" I wonder aloud to him.

> Absolutely I think it's possible if we imagine a world where these families, their young people, are the experts on their own lived experience—and that we create better pathways and opportunities. For folks to get involved in various forms of technology creation and dissemination, we need more diverse teams, we need teams that honor lived experiences, we need teams that can think beyond capitalism. I think capitalism is not going anywhere, but we need folks that can help induce ideas of humanity and anticipate challenges and issues. And that's going to take a lot of work, but there are lots of amazing, brilliant people that are working on these issues. A lot of tech companies should really think about who, who is leading these companies. Because I've encountered lots of really great folks in tech companies; That they have the same goals that I have but oftentimes they are mid-level folks, and so we can work together—they can bring me on as a consultant, they can bring other people on as consultants, but they can only go so far. They can only have so much power. So what I haven't seen tech companies do is, like, you know, actually—you might need a new CEO. We need a new C-suite that looks like America. I haven't seen that happen yet.
>
> —Dr. Desmond Patton[20]

The zone of proximal development that explains children's learning can also help us understand why so many with power and privilege to change the forecast observe either no reason for such changes (a cultural blindness) or present no will for the work that would result in change; despair kills social reform and those with interests to expand TSM's reach and endurance can create the very conditions that spur despair. An example is all too poignantly on display as testimony in US Senate subcommittees reveals that indeed adolescents who engage more hours of TSM report less self-compassion, greater mental health symptoms, and greater social distrust. These are the despairing precursors that help explain the increased suicide rate in our youth.

The self-regulatory impact another human body has on us should be obvious by now after this series of links to the Touchpoints model of infant mental health; proximity to caregivers is a human attachment function and objective. We function better, generally, with other bodies around us. As a result, many TSM users moderate other users. And these efforts, too, are then used to vilify—not the industry or the specific institutions, but rather the entire concept of community care. Labeled "cancel culture," online individual efforts to moderate and mediate comments, from the comments of the game industry's sexism, racism, classism, and colonialism to the comments of public officials called on to consider the community being attacked in their generalized statements, are maligned as the worst example of mob culture, when they represent an important and increasingly essential check on the myth of progress and social deliberation. Individual liberties are notions of free expression that favor freely expressed denigration of unprotected groups but that do little to amplify the free expression of groups marginalized within and outside of TSM spaces. Yet this is the reality that many children whose cultural and class experiences are far from the middle class, educated, white conception of liberty and accountability. These children are now subjected to laws limiting their access to knowledge because of the threat such realism presents to the idea of whiteness that otherwise has colored their classrooms.

Adults, and companies, too, have ZPD, and for most they will fall outside the capacities they might otherwise be expected to have by virtue of their lack of appreciation and understanding of how such social realities are experienced within a given TSM. When voice activation is built into the AI capacity of a device like "Alexa," white American English speakers without "accent" can engage, but they can't conceive and then design for the accent variation available not only in the world, but even in the region just outside the office where such planning and programming was executed. The same is true for those technologists creating from their experience, but doing so in contexts with little experience accepting non-dominant, non-white, innovators. As much as Dr. Patton and I revel in the vision of TSM being shaped within communities and by those whose backgrounds look more like those of ancestors harvesting sugar rather than innovating for space exploration, we also know what it is to be the first, the only, or the "one" in a group of thinkers and scholars desiring difference but whose ZPD lacks attention to power, difference, and exclusion. It's an experience Dr. Kristopher Alexander describes in our interview exploring gaming and education in these contexts of social justice and economic inclusion.

> Like, I know I'm one of the first. I call myself a Black faculty member at the school in 70 years. And I know that I'm not the right type of Black because they don't usually talk about me very much, because wh— who wants to talk about the present Black father who's technologically competent, who help pay his parents mortgage? That's not in the news. There's no shows or video games that even talk about this. Like, it's like . . . And I said this in another interview and made somebody very upset. It's like I don't exist and I'm okay with not existing.
>
> —Dr. Kristopher Alexander, Professor of
> Video Game Design, Ryerson University[21]

We all lose when we allow regulation conversations to be about individual parents or children, we all lose when TSM is dictated by districts without care for family systems or history denied and disrupted by TSM, and we all lose when regulation in the context of individual tech companies is allowed to be controlled as though user moderation and critique of individual TSM (cancel culture) is used to distract parents and politicians from the levers of regulation available in the administrative state designed for the express purpose of positioning and promoting public health and free expression while holding the public interest and trust in their regulatory demands. Human attachment is not unidirectional, nor is it so sensitive it can't continue in the face of critique; indeed, if an infant simply stopped attempting to connect with a parent, there would be no further development possible. Instead ruptures in infant and parental attachment are negotiated actively by the infant and the caregiver, allowing both to find a way back to harmonious relationship. In the context of TSM, this could look like explaining, affirming, and reflecting, but in the environment of TSM spaces there is little built into the tools that incentivize such processes.

The attention to call out or cancel culture allows politicians to not engage with actual policy changes that could curtail TSM monopolies, reverse environmental consequences of endless innovation cycles that do not attend to the poisons of e-waste or the privations of those who collect it second hand or mine it to begin with. Focus on the freedoms of fascists to spread their message is critical, but doing so allows leaders to not consider the rights and freedoms of users to be free from surveillance, marketization, and extraction. The invisible hand of progress is not so invisible when we consider the history, people, forces, and interactions that bring us to this precipice, nor is it when we understand the overarching narratives that have pushed nation-building into earth destruction in the name of empire and capital control. It may be true that human economic development has improved humans' length and quality of life, but it is also true that this improvement is not evenly experienced, is not sustained without human privation, and is not an exclusively human endeavor or impact. Our feral ecologies are ever-expanding and their effects, as we navigate through ever more COVID-19 variations and surges, cannot be ignored.

This brings us back to hope, and despair, and the societal nexus that unregulated TSM has brought to us. Hope is the fuel of progress and resistance. TSM's ability to allow free expression and free enterprise is also its ability to control, contort, and constrain freedom, both in the hands of totalitarian states, in the actions of monopolistic and monopsonist market contexts, and in the individual networks where children and their parents interact with it. Will we be held hostage to the militarization of technology inherited from humans reaching for space and the defense industry, or to the creeping social isolation of the 1970s programmers, or to the corporatization of technology innovation of the 1990s, or perhaps instead now to the dystopian dreams of the venture capitalists angel funding the apps that will keep our next three generations with augmented worlds they alone control and experience? Is there any other way for technology ethics to find a place in technology development outside of post-hoc hand wringing? Are we culture thinkers and current communities only able to influence technology when it finds a market position, or is there a way to deepen the thinking, widen the view, elongate the timeline?

• *15* •

Memory, Creativity, Courage

Three questions opened this book and prompted my attention and research on TSM, and they were all about my concerns for my children and the lingering impacts TSM would have on their brains, their lives, and their future opportunities. I thought my research would focus on the studies related to mental health, identity, addiction, media studies, personality, and, of course, human development. What I hadn't realized, either in the intentional research work I conducted, or in the incidental research I would engage by living through rapid technological and ecological change, was how empty my questions were once I drilled further into the impacts and implications of the TSM in our lives. I wasn't prepared for the interconnections between space and TSM, the ways in which current community infrastructure issues were related to robber barons of the 1800s, or how the critiques around TSM use in children or by adults were so often harbingers of political and cultural power grabs as much as they might be about informed users.

What was happening in my kitchen, my professional life, my community economy, and my social circles was happening globally, and it was being shaped internal and external to my familiar social and spatial bubbles. I wasn't alone, which certainly is relieving, but neither was I but an incidental blip in a centuries-long battle between industry and government and increasingly between government and government. I thought I was going to learn more to protect myself from hate lists and tighten the parental controls on my kids. The "Make Way for Ducklings" desire of order, stability, community support, and family endurance seemed to me a proper ending for the research threads examined based on a foundation of human attachment and family support through anticipatory guidance. If enough families understood that the odds are stacked against our control, we might work to better communicate the cultural values and social good we want for our kids. *We just gotta connect with our kids more fully,* I thought during my beachside conversation with Ben's grandparents and my California dinner with Lyon.

A vision of man on the moon and of the promises of an integrated mountaintop spurred a generation toward the conditions that enabled massive explosion of telecommunication's innovations, organized Black communities globally against oppression, and made possible global space exploration. The fears of men from the moon ushered in nuclear disarmament and ended the controls of totalitarian regimes across the continent of Europe and Asia as the USSR fell. Today, as the levers of power at my local and national levels lead choruses of constituents against teaching accurate history in our resegregated public schools, there is another turning of the tide. And wealthy men launching space junk are themselves through the largess enabled by their extraction providing the telecommunications resources that are attempting to stop the encroachment of Russia into Ukraine, and beyond. Of course, in the 1960s and 1990s neither the US moon shot or the gift of duck statues could alter the geopolitical machinations within and between nations, but each shifted the paths previously possible.

We are at a moment that cannot be underestimated in terms of its paths and the shifts possible. But it would be unrealistic to think seeing interconnections alone would shift anything. We are equally unrealistic if we believe TSM is about parenting or specific TSM policymaking in our schools or homes. We must appreciate the full picture of the attachment disruptions that occur through TSM, the attachment repair they make possible, and recognize in them both a ray of hope to inspire and instill the kind of actions that could really help repair the ruptures this book describes.

Once I picked at the edges of how hate groups operate, how algorithms were conceived, what impact positive parasocial engagement can have on a brain primed for social cues, and once I realized how value-laden even regulatory opportunities are, it was obvious that even intentional parenting around TSM would have its limits, and that these limits were not so much about malice or bad actors in their own individual dramas, but about the corruption of power dynamics that control the legal and economic frameworks from which our civilization operates. I had not realized that I would be decolonizing technology or applying findings from legal research of critical race theory to help understand the severe limits of parents' abilities in supporting their children's engagement with TSM. And yet, increasingly, the journey led to paths with unexpected starting places, and unacceptable ending trajectories.

Technology has a way of transforming, as the 1967 film *The Graduate*'s most famous line, "Are you listening? Plastics," implied. The line, said to the soon-to-be college graduate about the industry he should be entering, was a symbol of the disconnect between the confused ambivalence of the youthful character, and the confident, if cryptic, ambitions of his father's business associate. Today, if that same scene were to play out, we might hear him say, "Are you listening? Graphene"![1] instead of plastics, as the two-dimensional material science breakout has come to revolutionize everything from touchscreens, batteries, human organs, and now even space exploration as it is being considered in the construction of the long-anticipated "space elevator." If I am being honest, with myself and with you the reader, the questions these technology touchpoints invite are less about the developmental moment my child, or as discussed in chapter 11, even myself, might be navigating. They are more about, or maybe they are *also* about, the developmental moment of my family, my workplace, my community, and our broader globe.

Just as human growth and development has predictable periods of growth, regression, consolidation, and adaptation, the growth of human systems does as well. Technology is a human system, and technology's touchpoints also have predictable periods of growth, regression, consolidation, and adaptation. The cycle of technology innovation has a disruptive impact on human growth and development, and while humans may gain a capital good through its development, their own development is altered by its introduction and adoption. Therefore, we must attend to the development cycles of humans, our technology, and our ecology by appreciating the bioethical imperative. Bioethics is the study of the effects of advances in biology, medicine, and technologies, and it rests on four key principles: non-malfeasance, beneficence, justice, and autonomy. It is both a timeless field, in that humans often seek these values in their respective eras, and very young, about fifty years or so. As with any ethics, there is debate about the comprehensiveness of these principles.[2]

A bioethical approach to the relationship between humans and our technology would consider the obligations human creators and human systems have to individual humans and to humanity. A chemist who actively understood the toxicity of a substance and then worked tirelessly to place the substance into every child and worker's body would rightfully be called evil once their knowledge or and commitment to harm was revealed. Bioethics[3] of technology development would consider that development is not agnostic or without cultural or historic precedent; that technology development leads to change and that change disproportionately impacts those with least power and access. A technology bioethics would consider what is possible, what is meaningful, and what is leverageable between what one market wants and another community needs.

Yet it's hard to know how such a responsive set of solutions could be created in a world where individual corporate actors can ignite a new technology on the backs of existing infrastructure, especially when such efforts are enabled to actively derail existing public good efforts. Nevermind imagining how such a process would unfold while sustaining our ecosystem and by releasing a multinational corporate hold on people and places. In some ways we see how it might be possible in films such as *Coded Bias*, and yet even as that film led to changes in public opinion, there were specific experts located within technology companies whose work to fix the problems of algorithmic bias was simultaneously squelched as they were forced out of the companies building these technologies. Is the public groundswell enough to bring forth technologies informed by bioethics? In my interview with SAFE lab director Dr. Desmond Patton, he sees the conversations having matured in the aftermath of George Floyd's killing and the protests against anti-blackness that followed. He shares that:

> It's been really clear that folks want to talk about AI and ethics folks want to talk about diversity in AI and that has just really increased over the last nine years. Folks initially were interested in kind of the sensational aspects of using AI to focus on violence and to predict. Over the years, what that has become as folks being more interested in the methodology that I use the ways in which I work with young people; social work perspectives social work thinking so I get a lot of opportunities to talk about my approach and how social workers inform how and why I do the work that I do.[4]

When Dr. Patton describes interest in the methods he is using, he means his work as a social worker, and his model of considering trauma, community context, and explicit ways to empower community members who are outside the ivory tower or ed tech environments. That type of community care and contextualization is rare for those used to studying computing or policing, and so his approaches are understood as valuable and unique in the context of solutions that have removed actual people from the equation. There is a reason his approaches work better; they are anchored in human realities, and it is these realities that we should be invested in understanding if we are to cocreate TSM regulations that will benefit not only the ideal user who is most like the original programmer, but actual users who will likely be limited in what they can adjust, locate, or limit when they interact daily with whatever TSM innovation is next disrupting their day.

Where should this lead us now? What is the anticipatory guidance that can be offered as my children crest their next developmental wave, as I begin another developmental chapter of my own, as our nation and indeed the world in various committee chambers consider the best next course for our collective societal development to take, with so many dashboard lights alerting us to the increasing obstacles to this end? What is the advice now? We might consider the multiple ways that TSM can shape and alter human development, the various technology touchpoints and all the ruptures that seem impervious to repair and think there is little to do but disconnect. Yet this response is not the one advocated by one of Fred Rogers's proteges. When Chip Donohue and I explore the implications of the emerging Senate subcommittee findings in the context of remote learning that is still a reality, he urges me to reconsider the Luddite options of fighting against technology in classrooms. To me or others he hears arguing against consistent TSM access in classrooms, he says:

> "Oh, oh, we've got to get together again, we finally can be in the classroom together. Relationships are first." Um, yes, relationships are first. I've, I've spent . . . Since 2009, I've been a senior fellow at the Fred Rogers Center. Of course, relationships matter more. The technology's just a way of supporting that.[5]

Supporting relationships is the point of TSM, at least for its best use with children. When it comes to adults, it may or may not be a tool that can assist in such lofty aims, and yet it is built on the foundation of being a community good, which all infrastructure is since the earliest days of transportation and telecommunication. It's ability to support commerce, economic growth, and markets is what enabled it to be developed to begin with. Now as entire new economies are built, as the impacts of our feral ecologies yield COVID remote realities, it is naïve to pretend TSM won't be in every child's life, directly through their own use, and politically through the control and manipulations of the governments and the actors who are bent to control both economic and information possibilities for others. As Americans retreat from the fray of political changes demanded in protests in 2020 and beyond, it is all too apparent that even more transformation and extraction is planned as telecommunication prepares for intergalactic transmission. The creation of Space Force, the sixth armed services branch established in 2019, and the development and reporting of national mineral deposits[6] are two infrastructures tied to energy and space travel that portend the innovation demand that will next be commercialized and grown into everyday life.

In the year or more leading up to the 2020 election, national headlines were reporting on all sorts of symptoms of dysregulation. "White Genocide" banners appeared on the lawn of my children's elementary school, and construction crews bedecked with Confederate flags, gunracks, and Texas bullhorns staked out local intersections sending intimidating signals to drivers around the corner from the same school, also our local polling station. I had endured another year of intense workplace discrimination and the emotional labor of coaching others living through theirs as a minoritized woman in a state low in numbers of professionally credentialed Black and Latina women. To say my vigilance was heightened, my emotions dysregulated, would be accurate.

I was more than ready for the sabbatical year that promised no teaching, administration, mentoring, coaching, or service of any kind.[7] Despite this, I was prepared to "go out" on my sabbatical with one final employer-facilitated request when what came through was to respond to mass violence brought forward by hate. In August 2019, just a day after a Trump campaign rally where the campaigning president demanded border walls and quoted "America First," a man with an AR-15 drove nearly one hundred miles to El Paso Texas, expressly to kill Latinx and Hispanic people whom he had decided didn't deserve to live. Could I comment as a clinical psychologist and explain why the night after the Texas shooting, hundreds of miles away from El Paso, thousands of people in New York City's Times Square were petrified when a backfiring car sent hundreds running in fear, believing they, too, were being targeted for being minority? When I get the request, I am more than dysregulated; I am outraged.

Instead of receiving comments from colleagues with expertise in political rhetoric and hate speech, the college-retained PR firm requests a mental health commentary. It seemed to me as if responding to the racialized social targeting unfolding with increasing intensity in an understandably reactive way was the healthy reaction, and that the abnormality requiring examination was the silent endorsement such queries enabled for the ongoing and resurgent terrorism that is bound to white supremacy. This is the power of white supremacy in regulating itself and the outcomes that result from silence, diversion, and subversion. And the hardest part was realizing that no one positioned with power in the institutions involved (my own, the PR firm, or the news cycle where such commentary would be placed) understood the implications of their outreach nor the insidiousness of the supremacy they enacted. The zone of proximal development for whiteness is a deep one,[8] one that I was not prepared to scaffold and support while attempting to meet my own emotional and psychological needs in the face of white supremacist terror playing out across the country in increasing rates.

I wrote back to the PR firm with a statement, one they refused to bring to press. I found in my outrage no comfort, and it, combined with other ongoing factors existing even before the pandemic, set in motion more than a year of alienation from my colleagues who seemed oblivious to the impact of the systemic dismissal and denial such practices repeated over a career might bring. I wrote to the PR firm director that

> A clinician *is* needed I suppose; there is certainly info to share about dealing with newly surfaced agoraphobic symptoms and there is some concrete advice to offer trauma survivors who feel reactivated by the horrors around us. Instead of linking policy discussions on gun control and public health as rooted in a specific white supremacist ethos we debate increased individual emotion regulation that might change community outcomes. When the neighborhoods changed in cities across the US, we called it "white flight," now that social media neighborhoods contain more opportunities for mixed conversations it's time for good white people to stop "white fright" from keeping them from using their voices against white supremacy.

It was a bit overwrought, as I reflect, and I understand now why the statement wasn't used by the firm, given both its lack of specificity and its generalizations about neighborhoods offline and online.

Looking back with greater insights into online neighborhoods, I can also appreciate how distinct experiences within TSM are for individuals, sorted algorithmically by race, age, gender, region, religion, pet ownership status, and all the myriad ways our own engagement and the advertising metrics create the online realities that we each navigate. It was part of why a group of New York City Latinx communities could startle and believe themselves as under threat as shoppers at an El Paso Walmart had been only days earlier; the connectivity we feel online to real groups is real in our bodies and minds, and no actual distance can reduce that reality as we form and re-form our sense of self and community through our daily engagements, online and off. Just as I could not fathom the blindness to upholding whiteness playing out by nice people doing their work, they (my colleagues, the PR firm) could not fathom the daily reality of knowing you and others like you are under unrelenting attack.

When headlines of hate killings unfold anywhere, when videos of traffic stops end in death or "suicide," when these realities across screens in minute-by-minute scrolls are the realities of so many minoritized, there is no event that is far enough removed from daily experience as to be a dispassionate event worthy of neutral emotionless commentary. At least, there shouldn't be. Any opportunity to bring forward awareness within the contact zones between communities that such media statements might afford feels both imperative yet imperfect, and neither is likely to leave the one minority being asked to speak feeling empowered. If I can take liberty with Audre Lorde's sentiments,[9] belonging cannot require the abandoning of our differences, but the expanding of our attachments.

We are earthbound, no matter how our stars are aligned. The innovation and the minerals and networks they require will be highly contested in the next half century, and our best hope is organizing around shared human and ecological well-being. We must confront the simplistic teleological story of progress and demand accountability for demonstrating human and ecological impacts *before* industrialization, and we cannot rely on political and business leaders alone to be experts in reform or regulation. Technology bioethics requires both historical understanding and future imaginings. It requires us to consider what is essential about humanness and to move beyond it to consider the ecology we require as humans; what is the best version of human life that can be remembered and observed, and that takes care to create and innovate TSM with this awareness and sensitivity in mind.

OUR CONNECTED WORLD

In 2011, when then Pope John Paul IX began the Twitter profile @pontifex, I was one of the first hundreds of thousands of followers; my college account followed and retweeted, and soon I added His Eminence to my feed. It's no surprise then that seven years later, when Pope Francis gave the first official papal TED Talk, again I was one of the first hundreds of thousands of viewers. The conference theme that year, The Future You, featured deep fifteen-minute dives into the ways innovation, design, and education would transform each viewer into a new, improved version of themselves. In the context of forward progress, his remarks about how the keys to tomorrow lie in our ability to have memory, creativity, and courage, but most importantly how all of this (when available and combined through human interaction) creates the greatest renewable energy source on the planet: human hope.

I have watched the talk many times since; it was a challenging message to receive, frankly. By 2017 (the time of his remarks), it was possible to retweet an injustice, share a Kickstarter campaign, and reconnect (or lose touch) intentionally with any variety of actual people I shared presence within life, and to instead rely on the technological projections of social media existence to suffice for relationship. Indeed, by that time my own non-household relationships felt strained enough in person that the screed was preferred to the need. The Pope was acknowledging the power of such an existence in connecting us across the globe and to providing even greater spiritual insights through such sharing experiences. His was a call for humans to be in solidarity with one another's pain, and, in doing so, to seize those moments as the most vital to our own continuation as individuals and as a society, to physically be with others. I never wanted a pandemic to force my internet-connected social existence exclusively online, but let's say that after the trauma of those years leading up to 2020, I was ready for the remote life.

Technology can allow connection, the Pope affirmed and forecasted this, and yet, it can allow for us to ignore the disconnection around and between connections as well. How would memory, creativity, and courage overcome that truth? Was individual action and attention to one singular other really the key to unwinding this disconnected tangle of our time? As I review the research on attachment disruption and repair, what comes up is not so different from what the Pope in his remarks dictated: Narrative coherence, meaning, and purpose are the ingredients consistently heralded as essential to healthy human relations. Understanding our own story, creating a sense of coherence, and using that to drive and support individual conceptualizations of one's purpose in living are the features that extend from secure attachments and that can be built upon through treatment for those who are disconnected. When these are present or built anew in a person, they can achieve personally and socially meaningful and impactful things. When they are absent, disruption is not only the root, but the consequence.

We are beginning to see the ripple effects of memory, creativity, and courage aligning as individual actions yield socially coherent narratives that help us collectively find meaning and resolve a new purpose in our deliberations about TSM adoption and guidance. What I suspected as I fretted over whether my children were prepared to manage real life is finally coming public as whistleblower Frances Haugen testifies before US Senate and House committees· revealing herself and identity after the publication of sixty thousand documents she identified and archived in order to highlight the internal research affirming the corrosive, destabilizing, and dehumanizing impacts the platform of Facebook has had on individual children and on entire societies and ethnic groups.[10]

As Frances reveals her efforts to follow career decisions of colleagues to unearth their and others' comments relative to human trafficking functions of the platform, or suicidal ideation impacts of video recommendation algorithms, or contend with the years-long understanding of body image on young women using Instagram, I am struck by the personal memory, creativity, and courage it reveals. In accepting a job at Facebook, she had only one goal, one that grew from her observing the radicalization and isolation of a close white male friend. Once inside, and as she observed the platform cavalierly roll back protections used leading into the 2020 US presidential election, she reports having candid conversations with her mom. Her mom was not only Frances's mother, but also an ordained Episcopal priest. And just as Pope Francis demanded of his 2017 TED Talk audience, and frankly every future audience of the TED Talk, her mother challenged her:

If your one life and your one presence can bring hope to the world, can save a life, then that is what is asked of you in solidarity with humanity.

I am not naïve; having read Horwitz's account of the telecommunication deregulation patterns of the last century, I can recognize the various machinations at play. All the testimony and hearings would only be happening if there was political will to do more than hold hearings. It's a lot to see the response, the immediate hearings, the immediate support of an attractive young blond white woman, having also watched, supported, and continuously tuned in to the rejection of others who raised similar or equally startling alarms; women of color like Timnit Gebru at Google and Sophie Zhang at Facebook. It's a lot to see the machinations that will likely break big tech up and yet leave untouched core factors that enabled the harm already wrought. Clearly this whistleblower is the "right" one, and this presidential administration and pre-midterm moment is the "right" one to bring to bear political regulations that can impact these companies and practices far into the future. However clear these things may be in my own estimation, it is also clear that the TSM issues are not resolvable without addressing capital flow, regulation of markets, and creating structures that address social inequity. Still, the moves happening are important.

As Frances argues in her testimony before the Senate and House committees, she is not shy in naming these tensions herself. For whatever motivations and methods I might conjecture or critique to explain her choices leading up to her release of the materials, it's clear this is the fight she intends to land with, and whether or not there are long-term systemic changes, there will certainly be short-term changes to Facebook's power. What does it require then? What is it that we need, and what should be anticipated? Just like the work of considering the next developmental aim of our children, we do well to consider the developmental aim of our social world, because it is in considering this that we will find our best guidance with which to anticipate for ourselves and the TSM, and the children, not yet conceived.

It's clear that Europe and California will continue to be leaders in public interest policy protections, while China and the United States will continue to lead in AI arms battles across several essential domains, including military operations, medical decisions, and community policing. These realities mean there are identifiable ways to engage at the level of the lever, in hyperlocal and networked deliberative and iterative trials as adoption, investments, and impact are considered. And for many of the proposed changes, improvements, or overhauls, we must extend them with full awareness of the creep and reach of fascism, sexism, racism, and totalitarianism. Just as we ask engineers to "Black Mirror" their disruptive tech innovation, it's worth applying that lens to proposed policies.

And so, I try this for myself, imagining the worst of the outcomes of the ambivalent, preoccupied, dismissive, and avoidant parenting policies I have individually—and we have collectively—applied to our technology adoptions and those of our children. But the episode I envision feels as though it has come to life as a week of wounded men and boys wreaking havoc with vehicles and guns in safe places celebrating the return of normal, and a slew of courtrooms where women's trauma and pain and culpability for harm and heroism is also highly visible. It's hard not to understand there is a need for the attunement, responsiveness, repair, and attachment that our physiology is aligned with in our most intimate social relations. Indeed, whether it is Pope Francis's call for solidarity in action to foster hope as energy; or Frances Haugen's call for solidarity in action to foster transparency in understanding so as to counter the negative forces of Facebook and other TSM; whether it is the specifics of the techniques used by China and damned in Senate testimony; or whether it is the scientific consensus around what the answer is to address algorithmic bias, there is increasing consensus that what lies at the center of redressing TSM harm is the very thing that was supposed to be empowered by its free rein to begin with: individuals.

How to mitigate biased algorithms turned by faulty machine learning? Individuals will be needed to retrain the machines. Lots of them. Frequently, and in pace with rapidly shifting social frequencies. Predictions are now being held out that young adolescent women will increasingly fill this role. Individuals will be needed to deliberatively moderate decisions made by algorithms or other individuals as well; judges and moderators will have increasing power and decreasing visibility, resulting in the need for even more individuals: individuals to absorb the non-machine work of making machines learn and work and to process the emotional and physical needs of the humans making the machines and humans work better.

France/is (Haugen or Pope) is right to call for the power of the person to effect impact and to save lives. As much as this book is about parenting and policymaking, it's also about power. When I support a family in anticipating their child's next milestone and regression, I am preparing them to anticipate, plan around, and preemptively reframe the predicted human response to relational disruption combined with individual circadian and hormonal dysregulation. I know the supported parent provides the context for the child to be supported. Socially then, we must extend this knowledge to what our government and governance structures need to be supported even as their individual leaders and members are reaching milestones or causing disruption in the forward movement of our social world. We need to extend this to what our individual technology leaders need as they seek to leverage their narrative, its meaning, and their purpose. Just like Frances and Francis called for individuals, they both also called for connection.

When Frances Haugen during Senate testimony is asked what it takes for others to come forward to protect humanity from technology's capital-driven machinations, she becomes pensive. She relates the isolation that being a clear voice against institutional power brings. She identifies that others before her knew and saw the same things she knew and saw, but there was not an environment where that knowing and seeing would lead to anything but disconnection. Humans seek attachment, and systems that attach humans hold an enormous amount of power. Whether it's the power of the platform, or the power of an employer, sounding alarms in the service of these bioethical demands must not mean exile for those with the courage to raise them. And yet, in Frances's comments, we recognize that, too, is the fate she knew and yielded to in order to fulfill her purpose. We may develop more responsive TSM when average people can have courage, but they must believe there will still be a place for them to perform and belong professionally, rather than the isolation and alienation that befalls whistleblowers normally.

Patricia Greenfield writes about the mutual development that occurs between humans, technology, and within historical time periods. She frames ecological shifts in terms of two German words indicating "community" (*Gemeinschaft*, low-technology environments) and "society" (*Gesellschaft*, high-technology environments, also discussed in Chapter 10). There is a transmutation that occurs with and through transformation when it comes to the shifts made between generations, shifts that increasingly are devised and delivered through TSM. In her work, she delineates ways in which the technological and sociological transitions of a time and place represent influences on human development, and human development itself creates coevolutionary impact on its time and place such that economies and individual organization trajectories are intimately tied to the patterns that emerge because of the coevolution. That this particular inflection point arises with TSM represents a profound co-revolution on our society at a time when socially our predominance of adolescence, the cohort with the largest amount of social transformation energy, and the one with the greatest potential to steward our natural resources and care for their own well-being while tending aging populations and emerging populations behind them.

Perhaps it's unfair to conclude a volume centering human attachment through a metaphorical dialogue between Frances and Francis, when a central thesis was one that argued racialized policies, experiences, and expectancies are influencing TSM adoption with disparate impacts on vulnerable individuals, families, communities, economies, and ecologies. These two voices, from Western, developed places and their concurrent ideologies, represent insufficiently to position a solution created by exclusion. There are other voices and perspectives that are truly needed if we are to reset the relationship between our families and TSM. Those perspectives won't come through sanctioned channels, or comments that can be bot loaded to misrepresent public support. Our approach will need to be more input-rich, integrative, and iterative, with ruptures in impact carefully repaired in order to organize our best response to the dynamic changes.

Black Feminist Thought author Patricia Hill Collins, whose early work set the stage for the dynamic deepening of Black women's intellectual and cultural output across my college and grad school years, has more recent work engaging the ways that democracy and media interact. Hers was work that prompted Dr. Kishonna Gray's studies of Black women gamers, and it's not a mistake then to see in her work the crucial adaptation needed to best defend against inhumane TSM. She suggests that if both democracy and media are to be preserved, there will need to be flexible solidarities built through and across community. We do this, she offers, through intellectual politics, cultural politics, survival politics, and most especially with intersectional politics, building flexible solidarity whereby we find common interests to establish enough presence and pressure for voice on whatever issues are needing redress.

Second, intersectionality's focus on intersecting power relations suggests that prevailing theories of power and politics are far less universal than imagined. Neither liberalism, with its valorization of individual rights, and nor participatory democracy as a philosophy of how citizenship should work to ensure equality were designed with subordinated populations in mind. Blacks, women, ethnic groups, and similarly subordinated groups often served as markers for the absence of rights that defined citizenship. Political theory that relies on assumptions of an imagined, ideal and normative citizen may seem universalistic.[11]

When priorities of groups are not aligned, the solidarity can shift, but the function of writing, storytelling, community building, and organizing are all valued and necessary components of sustaining awareness across group identity zones of proximal development, so easily misused when weaponized to create division, and so powerfully bridged when recognized as paths toward solidarity. Serena Dokuaa Oduro phrases the crux of this lyrically as a poem, encapsulating the ethos of the Black Women Best framework, advocated by the US Department of Labor[12] when considering occupational health and safety of Black women across an economy. She writes:

> Do we need AI or
>   Do we need Black feminisms?
>   Liberation should lead,
>   Technology should support.
>   There is no mechanical solution
>   To sin.
>   There is only
>   The purposeful striving towards
>   Justice[13]

We should put our phones down and gather our children, of course, but not to swear off screens. Rather we should be using the screens to rise in flexible solidarity to resist the harm that industry and governments—empowered by infrastructure, conflict, and space exploration—would have us ignore as we fret over memes and moments. This ending is perhaps too expansive for a time that began with a mom, her kids, and some rotten community members. But every day, that is the substance of the TSM battles being fought. They don't all look like mine or my family's, yet they are all rooted in the same histories and histographies, and it is our hope that can bring us beyond the individual concerns we have over TSM in our homes and begin to help us attend and attune to the bigger patterns that are driving the wedges between us and our kids, if not their screens. We won't achieve this exclusively online, within one family, or even within a border. No resource exists in isolation or detached from the ecosystem that humans inhabit. The resources they rely on for the innovation they create are bound to the well-being of the planet and its resources, true even for those future humans whose existence will be space stations and eco colonies. Humans will always seek attachment, or to manage the disruptions on their own, and we will always be earth-bound. As TSM promises, we must also probe so that we can sustain the network in our collective and individual communities. We may not always find every overlap in interests, humor, or events when interacting in such an urgent time of response, but it's worth seeking. It's bigger, and smaller, than we can afford not to see any further.

Notes

CHAPTER 1

1. Bedingfield, Will, "The Far Right Are Running Riot on Steam and Discord," *Wired.* December 2021. https://www.wired.co.uk/article/steam -discord-far-right?utm_source=twitter&utm_medium=social&utm _campaign=onsite-share&utm_brand=wired-uk&utm_social-type=earned.

2. Musgrave, Shawn, "How White Nationalists Fooled the Media about Florida Shooter," Politico.com, February 2018. https://www.politico.com/story /2018/02/16/florida-shooting-white-nationalists-415672.

3. Kabali, Hilda K., Matilde M. Irigoyen, Rosemary Nunez-Davis, Jennifer G. Budacki, Sweta H. Mohanty, Kristin P. Leister, and Robert L. Bonner, "Exposure and Use of Mobile Media Devices by Young Children," *Pediatrics 136,* no. 6 (December 2015): 1044–50. DOI:10.1542/peds.2015-2151.

4. Choney, Suzanne, "Apple Offers iTunes Credits to Parents for In-App Purchases Made by Kids," June 24, 2013. https://www.nbcnews.com/tech /mobile/apple-offers-itunes-credits-parents-app-purchases-made-kids -f6C10424015.

5. Hirsh-Pasek, Kathy, Jennifer M. Zosh, Roberta Michnick Golinkoff, James H. Gray, Michael B. Robb, and Jordy Kaufman, "Putting Education in 'Educational' Apps: Lessons from the Science of Learning," *Association for Psychological Science 16,* no. 1 (April 2015): 3–34. DOI: 10.1177/1529100615569721.

6. National Association of Education for Young Children, *Technology and Interactive Media as Tools in Early Childhood Programs Serving Children from Birth through Age 8,* 2012. https://www.naeyc.org/sites/default/files/globally-shared /downloads/PDFs/resources/position-statements/ps_technology.pdf.

7. Guernsey, Lisa. *Screen Time: How Electronic Media—From Baby Videos to Educational Software—Affects Your Young Child* (New York: Basic Books, 2012).

8. Ibid.

9. Ibid.

10. Brazelton, T. Berry, and Joshua D. Sparrow. *Touchpoints—Birth to Three: Your Child's Emotional and Behavioral Development,* 2nd edition (Cambridge, MA: Da Capo Lifelong Books, 2006), 373.

CHAPTER 2

1. Tronick, Ed, and Marjorie Beeghly, "Infants' Meaning-Making and the Development of Mental Health Problems," *American Psychologist 66,* no. 2 (Winter, 2011): 107–19. DOI: 10.1037/a0021631.

2. Ibid.

3. Brazelton, T. Berry, and Joshua D. Sparrow, *Touchpoints—Birth to Three: Your Child's Emotional and Behavioral Development,* 2nd edition (Cambridge, MA: Da Capo Lifelong Books, 2006), 373.

4. "Brazelton Touchpoint Center," Brazelton Touchpoint Center. https://www.brazeltontouchpoints.org/.

5. Warner, Judith, *Perfect Madness: Motherhood in the Age of Anxiety* (New York: Riverhead Books, 2006).

6. Belkin, Lisa, "The Newest, Latest Parenting Trend," *Motherlode Blog,* November 8, 2010. https://parenting.blogs.nytimes.com/2010/11/08/the-newest-latest-parenting-trend/.

7. Brady, Loretta L.C., Rebecca Hadley, and Cathy Kuhn, "Creating a Family-Centered Wellness Team: Lessons Learned in Creating and Integrated Continuum of Care for Families Facing Homelessness, Addiction, and Trauma Recovery," *Journal of Social Distress and Homelessness 19,* no. 1–2 (July 2013): 83–106. DOI: 10.1179/105307809805365163.

8. Note: I use the abbreviation *TSM* rather than *T* or *SM* because while it is more common parlance to refer to Tech or Social Media, a central observation of this work is the way in which the unification and simultaneous emergence and proliferation of TSM is interconnected. Some scholars refer to this as Telecommunication Media (TCM); however, a psychological construct heavily explored in this text is Theory of Mind (TOM) and therefore this TSM convention is intended to ease the reader's distinguishing. Facebook doesn't track increasingly greater clicks per day without the ubiquitous access smartphones enabled. Facebook doesn't seek to create more channels for itself in acquiring Instagram without realizing that screentime per day doesn't fluctuate, but time on Facebook does as novelty decreases and complexity increases. So as this book considers the reflection point we are located within, it uses this term to acknowledge the need for intertechnology considerations.

9. *Protecting Kids Online: Snapchat, TIkTok, and YouTube*, October 2021, written testimony of Leslie Miller, vice president of YouTube government affairs and public policy. https://www.commerce.senate.gov/services /files/2FBF8DE5-9C3F-4974-87EE-01CB2D262EEA.

10. Calvert, Sandra L., and Melissa N. Richards, "Children's Parasocial Relationships," from *Media and the Well-being of Children and Adolescents*, A. Jordon and D. Romer, editors (New York: Oxford University Press, 2014), 187–200.

CHAPTER 3

1. Browne, Angela, Brenda Miller, and Eugene Maguin, "Prevalence and Severity of Lifetime Physical and Sexual Victimization Among Incarcerated Women," *International Journal of Law and Psychiatry* 22, no. 3–4 (1999): 301–322.

2. Brady, L.L.C., and E.P. Ossoff, "Fifty Years of Fashion and Feminism: The Effect of Early Recognition on Vocational and Sociopolitical Identity Development," *Current Psychology, 29*, no. 10 (2010): 34–44.

3. Wellman, Henry M., and Karen Lind. *Reading Minds: How Childhood Teaches Us to Understand People* (New York: Oxford University Press, 2020).

4. Nugent, J. Kevin, Constance H. Keefer, Susan Minear, Lise C. Johnson, and Yvette Blanchard, *Understanding Newborn Behavior and Early Relationships: The Newborn Behavioral Observations (NBO) System Handbook* (Baltimore: Paul H. Brookes, 2007).

5. Ibid.

6. Dawkins, Richard, "The Selfish Meme," *Time 153*, no. 15 (1999): 52–53.

7. Brodie, Richard, *Virus of the Mind: The New Science of the Meme* (Carlsbad, CA: Hay House, Inc, 2011).

8. Gal, Noam, Limor Shifman, and Zohar Kampf, "'It Gets Better': Internet Memes and the Construction of Collective Identity," *New Media and Society 18*, no. 8 (2016): 1698–714.

9. Ibid., 1700.

10. Hern, Alex, "Cambridge Analytica: How Did It Turn Clicks into Votes?" *The Guardian*, May 6, 2018. https://www.theguardian.com/news/2018/may /06/cambridge-analytica-how-turn-clicks-into-votes-christopher-wylie.

11. *Hearing Before the United States Senate Committee on the Judiciary and the United State Senate Committee on Commerce, Science, and Transportation*, 2018, statement of Mark Zuckerberg, chairman and chief executive officer of Facebook. https://www.judiciary.senate.gov/imo/media/doc/04-10-18%20Zucker berg%20Testimony.pdf.

12. Confessore, Nicholas, "Cambridge Analytica and Facebook: The Scandal and the Fallout So Far," *New York Times*, April 4, 2018. https://www.nytimes.com/2018/04/04/us/politics/cambridge-analytica-scandal-fallout.html.

13. Bruner, Jerome, *Acts of Meaning* (Boston: Harvard University Press, 1990).

14. Rotter, Julian B., "Social Learning Theory," *Expectations and Actions: Expectancy-Value Models in Psychology 395* (1982).

15. Collins, Patricia Hill, *Another Kind of Public Education: Race, Schools, the Media, and Democratic Possibilities* (Boston: Beacon Press, 2009).

16. Zialcita, Paolo, "Facebook Pays $643,000 Fine for Fine in Cambridge Analytica Scandal," *NPR*, October 2019. https://www.npr.org/2019/10/30/774749376/facebook-pays-643-000-fine-for-role-in-cambridge-analytica-scandal.

17. Xavier, Jean, Julien Magnat, Alain Sherman, Soizic Gauthier, David Cohen, and Laurence Chaby, "A Developmental and Clinical Perspective of Rhythmic Interpersonal Coordination: From Mimicry Towards the Interconnection of Minds," *Journal of Physiology-Paris 110*, 4-part B (November 2016): 420–6. DOI: 10.1016/j.physparis.2017.06.001.

18. Fitts, Alex Sobel, "Techs' Harassment Crisis Now an Arsenal of Smoking Guns," Backchannel, March 24, 2017. https://www.wired.com/2017/03/techs-harassment-crisis-now-has-an-arsenal-of-smoking-guns/.

19. Hampton, Rachelle, "The Black Feminists Who Saw the Alt-Right Threat Coming," *Slate*, April 2019. https://slate.com/technology/2019/04/black-feminists-alt-right-twitter-gamergate.html.

20. Mueller III, Robert S., "Report on the Investigation into Russian Interference in the 2016 Presidential Election," US Department of Justice, Vol. I. (March 2019). https://www.justice.gov/archives/sco/file/1373816/download.

CHAPTER 4

1. McNamee, Roger, "I Helped Create This Mess. Here's How to Fix It," January 28, 2019. https://time.com/magazine/us/5505429/january-28th-2019-vol-193-no-3-u-s/.

2. Fry, Hannah, *Hello World: How to Be Human in the Age of the Machine* (New York: W. W. Norton & Company, 2019).

3. Gorwa, R., R. Binns, and C. Katzenbach, "Algorithmic Content Moderation: Technical and Political Challenges in the Automation of Platform Governance," *Big Data & Society* (February 2020): 1–15. DOI: 10.1177/2053951719897945.

4. Addams, Jane, "The Devil Baby at Hull House," from *The Best American Essays of the Century*, Joyce Carol Oates and Robert Atwan, editors (Boston: Houghton Mifflin Company, 2001), 75–89.

5. Takeuchi, Lori, and Reed Stevens, "The New Coviewing: Designing for Learning through Joint Media Engagement," December 8, 2011, Report of the Jean Gantz Cooney Center. https://joanganzcooneycenter.org/publication /the-new-coviewing-designing-for-learning-through-joint-media-engagement/.

6. Smogorzewska, Joanna, Grzegorz Szumski, and Pawel Grygiel, "Theory of Mind Goes to School: Does Educational Environmental Influence the Development of Theory of Mind in Middle Childhood?" *PLoS ONE* 15, no. 8 (August 2020). DOI:10.1371/journal.pone.0237524.

7. Bruner, Jerome, *Acts of Meaning* (Boston: Harvard University Press, 1990).

8. Brashier, N. M., and D. L. Schacter, "Aging in an Era of Fake News," *Current Directions in Psychological Science*, 29, no. 3 (2020): 316–23. https://doi.org /10.1177/0963721420915872.

9. Gladwell, Malcolm, *The Tipping Point: How Little Things Can Make a Big Difference* (Boston: Little, Brown, 2006).

10. Ferguson, Amber, and Kyle Sweson, "Texas Family Says Teen Killed Himself in Macabre 'Blue Whale' Online Challenge That's Alarming Schools," *Washington Post* (July 2017). https://www.washingtonpost.com/news/morning -mix/wp/2017/07/11/texas-family-says-teen-killed-himself-in-macabre-blue -whale-online-challenge-thats-alarming-schools/.

11. Younger, Shannan. "Info Parents Need to Know About the Blue Whale Challenge," *Chicago Parent*, July 2017. https://www.chicagoparent.com/parenting /info-parents-need-to-know-about-the-blue-whale-challenge/.

12. Ghansah, Rachel Kaadzi, "A Most American Terrorist: The Making of Dylan Roof," *GQ*, August 2017. https://www.gq.com/story/dylann-roof -making-of-an-american-terrorist.

13. Dickinson, EJ, "What Is the Momo Challenge? Why Parents Are Freaking Out About This Terrifying 'Game'," *Rolling Stone*, February 2019. https://www.rollingstone.com/culture/culture-news/what-is-the-momo-chal lenge-800470/.

14. Barnett, Clive, "Technologies of Citizenship: Assembling Media Publics," from *Culture and Democracy: Media, Space, and Representation* (Edinburgh: Edinburgh University Press, 2003), 81–107.

15. Brady, L., and E. Evans, "Financial Market Shifts in the Era of Regulation, Deregulation, and Global Markets: How an Ethic of Care Lens Can Reveal and Mitigate Financial Market Risks and the Ethical Responsibility this Entails," Presented at the 1st Annual The Ethics of Business, Trade, & Global Governance: An Interdisciplinary Conference, December 1, 2018, New Castle, New Hampshire.

CHAPTER 5

1. Cromer, Jason A., Adrian J. Schembri, Brian T. Harel, and Paul Maruff, "The Nature and Rate of Cognitive Maturation from Late Childhood to Adulthood," *Frontiers in Psychology* (May 2015). DOI: 10.3389/fpsyg.2015.00704.

2. Hong, Yoo Rha, and Jae Sun Park, "Impact of Attachment, Temperament, and Parenting on Human Development, *Korean Journal of Pediatrics 55*, no. 12 (December 2012): 449–54. DOI:10.3345/kjp.2012.55.12.449.

3. Darwin, Charles, "A Biographical Sketch of an Infant," *Mind 2*, no. 7 (1877): 285–94.

4. Darwin, Charles, *On the Origin of Species by Means of Natural Selection, or Preservation of Favored Races in the Struggle for Life* (London: John Murray, 1859).

5. Ainsworth, Mary S., and John Bowlby, "An Ethological Approach to Personality Development," *American Psychologist 46*, no. 4 (1991): 333.

6. Hess, Eckhard Heinrich, *Imprinting: Early Experience and the Developmental Psychobiology of Attachment*, Foreword by Konrad Lorenz (New York: Van Nostrand Reinhold, 1973).

7. Hubel, David H., and Torsten N. Wiesel, "Receptive Fields, Binocular Interaction and Functional Architecture in the Cat's Visual Cortex," *Journal of Physiology 160*, no. 1 (1962): 106.

8. Curtiss, Susan, Victoria Fronkin, Stephen Krashen, David Rigler, and Marilyn Rigler, "The Linguistic Development of Genie," *Language 50*, no. 3 (September 1974): 528–54. DOI: 10.2307/412222.

9. McCloskey, Robert, *Make Way for Ducklings* (London: Penguin Young Readers Group, 1941).

10. Christensen, Brandon, "10 Worst Space Disasters in History," *Real Clear History*, March 2019. https://www.realclearhistory.com/articles/2019/03/08/10_worst_space_disasters_in_history_420.html.

11. History.com Editors, "Space Shuttle Challenger Disaster," *History*, November 2009. https://www.history.com/this-day-in-history/challenger-explodes.

12. "About." McNair Scholars, McNair Scholars Program, https://mcnairscholars.com/about/.

13. Maranto, Lauren, "Who Benefits from China's Cybersecurity Laws?" *Center for Strategic and International Studies* (June 2020). https://www.csis.org/blogs/new-perspectives-asia/who-benefits-chinas-cybersecurity-laws.

14. Baetjer, Howard, "Capital as Embodied Knowledge: Some Implications for the Theory of Economic Growth," *Review of Austrian Economics 13*, no. 2 (2000): 147–74.

15. Guernsey, Lisa, interview by Loretta L.C. Brady, Zoom interview, Manchester, New Hampshire, October 2021.

16. "Dust or Magic Fall Institute via Zoom," Dust or Magic. http://dustor magic.com/institute/.

17. Guernsey, Lisa, interview by Loretta L.C. Brady, Zoom interview, Manchester, New Hampshire, October 2021.

18. McNair, Carl, "Eyes on the Stars," *Storycorps.* https://storycorps.org /animation/eyes-on-the-stars/.

19. "President's Committee of Advisors on Science and Technology," *Panel on Educational Technology* (March 1997). https://clintonwhitehouse3.archives .gov/WH/EOP/OSTP/NSTC/PCAST/k-12ed.html.

20. Wartella, Ellen, "A Brief History of Children and Media Research," Workshop on Media Exposure and Early Childhood Development, presented at Northwestern University, January 25, 2018. https://www.nichd.nih.gov/sites /default/files/2018-03/WartellaHistChildMediaResearch.pdf, accessed September 19, 2021.

21. Baccarella, Christian V., Timm F. Wagner, Jan F. Kietzmann, and Ian P. McCarthy, "Ascertaining the Rise of the Dark Side of Social Media: The Role of Sensitization and Regulation," *European Management Journal 38*, no. 1 (February 2020): 3–6. DOI: 10.1016/j.emj.2019.12.011.

22. Kriss, Alexander, *The Gaming Mind: A New Psychology of Videogames and the Power of Play* (New York: The Experiment, 2020).

23. Solomon, Feliz, "What 'The Hunger Games' Three-Finger Salute Means to Protesters Across Asia." https://www.wsj.com/articles/what-the-hunger -games-three-finger-salute-means-to-protesters-across-asia-11613649604, accessed February 2021.

24. Baccarella, Christian V., Timm F. Wagner, Jan F. Kietzmann, and Ian P. McCarthy, "Ascertaining the Rise of the Dark Side of Social Media: The Role of Sensitization and Regulation," *European Management Journal* 38, no. 1 (February 2020): 3–6. DOI: 10.1016/j.emj.2019.12.011.

CHAPTER 6

1. "Media Education. American Academy of Pediatrics. Committee on Public Education," *Pediatrics 104*, 2, Pt 1 (August 1999): 341–3. PMID: 10429023.

2. Vaala, Sarah, Anna Ly, and Michael H. Levine, "Getting a Read on the App Stores: A Market Scan and Analysis of Children's Literacy Apps" (New York, NY: The Joan Ganz Cooney Center at Sesame Workshop, Fall 2015), 1–50. https://www.joanganzcooneycenter.org/wp-content/uploads/2015/12/jgcc _gettingaread.pdf.

3. Berkowitz, Talia, Marjorie W. Schaeffer, Erin A. Maloney, Lori Peterson, Courtney Gregor, Susan C. Levine, and Sian L. Beilock, "Math at Home Adds

Up to Achievement in School," *Science 350*, no. 6257 (October 2015), 196–8. DOI:1 0.1126/science.aac7427.

4. Thomas, Balmès, director, *Babies*. Universal Studios Home Entertainment, 2010. 1:19:00.

5. Graham, Ruth, "Baby's First Photo: The Unstoppable Rise of the Ultrasound Souvenir Industry," *Buzzfeed News* (September 2014). https://www.buzzfeednews.com/article/ruthgraham/sharing-ultrasound-photos-facebook-instagram.

6. Haughton, Ciaran, Mary Aijen, and Carly Cheevers, "Cyber Babies: The Impact of Emerging Technology on the Developing Infant," *Psychology Research 5*, no. 9 (September 2015): 504–18. DOI: 10.17265/2159-5542/2015.09.002.

7. "Understanding the Development of Attachment Bonds and Attachment Behavior over the Life Course," from Daniel P. Brown and David S. Elliott, *Attachment Disturbances in Adults: Treatment for Comprehensive Repair* (New York: W.W. Norton, 2016).

8. "Assessment of Adult Attachment," from Daniel P. Brown and David S. Elliott, *Attachment Disturbances in Adults: Treatment for Comprehensive Repair* (New York: W.W. Norton, 2016).

9. Leslie, Alan M., "Pretending and Believing: Issues in the Theory of ToMM," *Cognition 50* (1994): 211–38. DOI: 0010-277(93)00601-A.

10. Przybylski, Andrew K., and Netta Weinstein, "A Large-Scale Test of the Goldilocks Hypothesis: Quantifying the Relationship Between Digital Screen Use and the Mental Well-being of Adolescents," *Psychological Science 28*, no. 2 (2017): 204–7. DOI: 10.1177/0956797616678438.

11. Mullaney, Thomas S., Benjamin Peters, Mar Hicks, and Kavita Philip, eds., *Your Computer Is on Fire* (Cambridge, MA: MIT Press, 2021), 117–34, 313–62.

CHAPTER 7

1. Brazelton, T. Berry, and Joshua D. Sparrow, *Touchpoints—Birth to Three: Your Child's Emotional and Behavioral Development*, 2nd edition (Cambridge, MA: Da Capo Lifelong Books, 2006), 373.

2. Brazelton, T. Berry, and Joshua D. Sparrow, *Touchpoints—Birth to Three: Your Child's Emotional and Behavioral Development*, 2nd edition (Cambridge, MA: Da Capo Lifelong Books, 2006); Brazelton, T. Berry, and Joshua D. Sparrow, *Touchpoints—Three to Six: Your Child's Emotional and Behavioral Development* (Cambridge, MA: Da Capo Lifelong Books, 2001).

3. Tronick, Ed, and Marjorie Beeghly, "Infants' Meaning-Making and the Development of Mental Health Problems," *American Psychologist 66*, no. 2 (Winter 2011): 107–19. DOI: 10.1037/a0021631.

4. Arnheim, Rudolf, *Visual Thinking* (Berkeley: University of California Press, 1969).

5. Rose, Megan, "The Average Parent Shares Almost 1,500 Images of Their Child Online Before Their 5th Birthday," *Parent Zone.* https://parent zone.org.uk/article/average-parent-shares-almost-1500-images-their-child-on line-their-5th-birthday.

6. Graham, Ruth, "Baby's First Photo: The Unstoppable Rise of the Ultrasound Souvenir Industry," *Buzzfeed News* (September 2014). https://www .buzzfeednews.com/article/ruthgraham/sharing-ultrasound-photos-facebook -instagram.

7. Prince, Dana, "What About Place? Considering the Role of Physical Environment on Youth Imagining of Future Possible Selves," *Journal of Youth Studies* 17, no. 6 (2014): 697–716.e.

8. Brady, Loretta L.C., "Canaries in the Ethical Coal Mine? Case Vignettes and Empirical Findings for How Psychology Leaders Have Adopted Twitter," *Ethics and Behavior 26*, no. 2 (February 2016): 110–27. DOI:10.1080/1050842 2.2014.994064.

9. *Protecting Kids Online: Snapchat, TikTok, and YouTube*, October 2021 written testimony of Leslie Miller, vice president of YouTube government affairs and public policy. https://www.commerce.senate.gov/services/files /2FBF8DE5-9C3F-4974-87EE-01CB2D262EEA.

10. Hernandez, Javier C., and Albee Zhang, "90 Minutes a Day, until 10 P.M.: China Sets Rules for Young Gamers," *New York Times* (November 6, 2019). https://www.nytimes.com/2019/11/06/business/china-video-game -ban-young.html.

11. Amer, Yasmin. 2021. "Henrietta Lacks' Family Sues Biotech Company Profiting from Stolen Cells." *All Things Considered.* NPR WBUR October 13, 2021.

12. Tronick, Ed, and Marjorie Beeghly, "Infants' Meaning-Making and the Development of Mental Health Problems," *American Psychologist 66*, no. 2 (Winter 2011): 107–19. DOI: 10.1037/a0021631.

13. Auxier, Brooke, Monica Anderson, Andrew Perrin, and Erica Turner, "Parenting Children in the Age of Screens," Pew Research Center (July 2020). https://www.pewresearch.org/internet/2020/07/28/parenting-children-in -the-age-of-screens/.

14. McCaleb, Miriam, Patricia Champion, and Phillip J. Schluter, "Commentary: The Real (?) Effect of Smartphones Use on Parenting: A Commentary on Modecki et al. (2020)," *Journal of Child Psychology and Psychiatry 62*, no. 12 (2021): 1494–96. DOI: 10.1111/jcpp/13413.

15. "Media Education. American Academy of Pediatrics. Committee on Public Education," *Pediatrics 104*, no. 2, Pt 1 (August 1999): 341–3. PMID: 10429023.

16. Berkowitz, Talia, Marjorie W. Schaeffer, Erin A. Maloney, Lori Peterson, Courtney Gregor, Susan C. Levine, and Sian L. Beilock, "Math at Home Adds Up to Achievement in School," *Science 350*, no. 6257 (October 2015): 196–8. DOI:10.1126/science.aac7427.

17. Brazelton, T. Berry, and Joshua D. Sparrow, *Touchpoints—Three to Six: Your Child's Emotional and Behavioral Development* (Cambridge, MA: Da Capo Press, 2001), 341.

18. Ibid., 371.

19. Ibid., 342.

20. Donohue, Chip, interview by Loretta L.C. Brady, Zoom interview, Manchester, New Hampshire, October 2021.

CHAPTER 8

1. DeFord, Carolyn, "Our Bodies Are Just a Shell: A Mother's Wisdom on Life and Death," *Story Corps*. December 4, 2021. https://storycorps.org/stories/our-bodies-are-just-a-shell-a-mothers-wisdom-on-life-and-death/.

2. Xavier, Jean, Julien Magnat, Alain Sherman, Soizic Gauthier, David Cohen, and Laurence Chaby, "A Developmental and Clinical Perspective of Rhythmic Interpersonal Coordination: From Mimicry Towards the Interconnection of Minds," *Journal of Physiology—Paris 110*, four-part B (November 2016): 420–6. DOI: 10.1016/j.physparis.2017.06.001.

3. Wellman, Henry M., and Karen Lind, *Reading Minds: How Childhood Teaches Us to Understand People* (New York: Oxford University Press, 2020).

4. Brazelton, T. Berry, and Joshua D. Sparrow, *Touchpoints—Birth to Three: Your Child's Emotional and Behavioral Development*, 2nd Edition (Cambridge, MA: Da Capo Lifelong Books, 2006), 373.

5. Wellman, Henry M., and Karen Lind, *Reading Minds: How Childhood Teaches Us to Understand People* (New York: Oxford University Press, 2020).

6. "Understanding the Development of Attachment Bonds and Attachment Behavior Over the Life Course," from Daniel P. Brown and David S. Elliott, *Attachment Disturbances in Adults: Treatment for Comprehensive Repair* (New York: W. W. Norton, 2016), 75–102.

7. Ibid.

8. Wellman, Henry M., and Karen Lind, *Reading Minds: How Childhood Teaches Us to Understand People* (New York: Oxford University Press, 2020).

9. Baetjer, Howard, "Capital as Embodied Knowledge: Some Implications for the Theory of Economic Growth," *Review of Austrian Economics 13*, no. 2 (2000): 147–4.

10. *Protecting Kids Online: Snapchat, TikTok, and YouTube,* October 2021, written testimony of Leslie Miller, vice president of YouTube government affairs and public policy. https://www.commerce.senate.gov/services/files/2FBF8DE5-9C3F-4974-87EE-01CB2D262EEA.

11. Donohue, Chip, and Roberta Schomburg, "Technology and Interactive Media in Early Childhood Programs: What We've Learned from Five Years of Research, Policy, and Practice," *YC Young Children 72,* no. 4 (2017): 72–8.

12. Donohue, Chip, interview by Loretta L.C. Brady, Zoom interview, Manchester, New Hampshire, October 2021.

13. Brazelton, T. Berry, and Joshua D. Sparrow, *Touchpoints—Birth to Three: Your Child's Emotional and Behavioral Development,* 2nd Edition (Cambridge, MA: Da Capo Lifelong Books, 2006), 344.

14. Karlis, Nicole, "A Generation of Kids Has Used Social Media Their Whole Lives. Here's How It's Changing Them," *Salon* (February 2021). https://www.salon.com/2021/02/11/a-generation-of-kids-has-used-social-media-their-whole-lives-heres-how-its-changing-them/.

15. Bandura, Albert, Joan E. Grusec, and Frances L. Menlove, "Observational Learning as a Function of Symbolization and Incentive Set," *Child Development* (1966): 499–506.

16. Kamp, David, *Sunny Days: The Children's Television Revolution That Changed America* (New York: Simon & Schuster, 2020).

17. Ibid.

CHAPTER 9

1. Kamp, David, *Sunny Days: The Children's Television Revolution That Changed America* (New York: Simon & Schuster, 2020).

2. Wartella, Ellen, "A Brief History of Children and Media Research. Workshop on Media Exposure and Early Childhood Development," presented at Northwestern University, January 25, 2018. https://www.nichd.nih.gov/sites/default/files/2018-03/WartellaHistChildMediaResearch.pdf, accessed September 19, 2021.

3. Mah, V., Kandice, and E. Lee Ford-Jones, "Spotlight on Middle Childhood: Rejuvenating the 'Forgotten Years,'" *Pediatrics Child Health 17,* no. 2 (February 2012): 81–3. DOI: 10.1093/pch.17.2.81.

4. Smith, Leslie, "Piaget's Infancy Journal: Epistemological Issues," *Constructivist Foundations 14,* no. 1 (2018): 85–7.

5. Smogorzewska, Joanna, Grzegorz Szumski, and Pawel Grygiel, "Theory of Mind Goes to School: Does Educational Environmental Influence the De-

velopment of Theory of Mind in Middle Childhood?" *PLoS ONE 15*, no. 8 (August 2020). DOI:10.1371/journal.pone.0237524.

6. Mah, V., Kandice, and E. Lee Ford-Jones, "Spotlight on Middle Childhood: Rejuvenating the 'Forgotten Years,'" *Pediatrics Child Health 17*, no. 2 (February 2012): 81–3. DOI: 10.1093/pch.17.2.81.

7. "Michael Apted on the Incredible Journey of the 'Up' Series," *Film at Lincoln Center* (2019). https://www.filmlinc.org/daily/michael-apted-on-the-incredible-journey-of-the-up-series/.

8. Ibid.

9. Almond Paul, director, *Seven Up*, Granada Television, 1964 (London, UK).

10. DelGuidice, "Middle Childhood: An Evolutionary-Developmental Synthesis," from N. Halfon, C. Forrest, R. Lerner, and E. Fraustman, eds., *Handbook of Life Course Health Developmental* (November 2017): 95–107. DOI: 10.1007/978-3-319-47143-3_5.

11. "Marketing Food to Children and Adolescents: A Review of Industry Expenditures, Activities, and Self-Regulation: A Federal Trade Commission Report to Congress," Federal Trade Commision. https://www.ftc.gov/reports/marketing-food-children-adolescents-review-industry-expenditures-activities-self-regulation.

12. Dugan, Therese E., Reed Stevens, and Siri Mehus, "From Show, to Room, to World: A Cross-Context Investigation of How Children Learn from Media Programming," *International Society of the Learning Sciences 1* (June 2010): 992–9. DOI: 10.22318/icls2010.1.992.

13. Mishna, Faye, Michael Saini, and Steven Solomon, "Ongoing and Online: Children and Youth's Perceptions of Cyber Bullying," *Children and Youth Services Review 31*, no. 12 (2009): 1222–8.

14. Halfon, Neal, and Miles Hochestein, "Life Course Health Development: An Integrated Framework for Developing Health, Policy, and Research," *MilBack Quarterly 80*, no. 3 (September 2002): 433–79. DOI: 10.1111/1468-0009.00019.

15. Winnicott, D.W., "Transitional Objects and Transitional Phenomena," *Playing & Reality* (London: Tavistock Publications, 1971), 1–18.

16. Perrin, Andrew, "5 Facts About Americans and Video Games," Pew Research Center (September 2018). https://www.pewresearch.org/fact-tank/2018/09/17/5-facts-about-americans-and-video-games/.

17. DelGuidice, Marco, "Middle Childhood: An Evolutionary-Developmental Synthesis," from N. Halfon, C. Forrest, R. Lerner, and E. Fraustman, eds., *Handbook of Life Course Health Developmental* (November 2017): 95–107. DOI: 10.1007/978-3-319-47143-3_5.

CHAPTER 10

1. Brazelton, T. Berry, and Joshua D. Sparrow, *Touchpoints—Three to Six: Your Child's Emotional and Behavioral Development* (Cambridge, MA: Da Capo Press, 2001).

2. Alexander, Kristopher, interview by Loretta L.C. Brady, Zoom interview, Manchester, New Hampshire, November 2021.

3. TEDxTalks, "They Are Children: How Posts on Social Media Lead to Gang Violence," YouTube video, 12:19, May 15, 2017. https://www.youtube.com/watch?v=BmlvOGh7Spo.

4. Baird, A.A., "Understanding Adolescent Decision Making" (invited talk), Association for Psychological Science, Annual Meeting, May 2007, Chicago, Illinois.

5. Ibid.

6. Webster-Stratton, Carolyn, "Cross-Cultural Collaboration to Deliver the Incredible Years Parent Program," *The Incredible Years* (2006): 1–34. https://incredibleyears.com/wp-content/uploads/training-interpreters-deliver-cross-cultural-colaboration_06.pdf.

7. Guz, Samantha, "Reflections on Practice," blog series School Social Work Network, "The Stories We Tell Ourselves, Part 4: School Social Work and White Womanhood," November 2, 2021. https://schoolsocialwork.net/the-stories-we-tell-ourselves-part-4-school-social-work-and-white-womanhood/, accessed January 14, 2022.

8. Greenfield, Patricia M., "Linking Social Change and Developmental Change: Shifting Pathways of Human Development," *Developmental Psychology 45*, no. 2 (2009): 401–18. DOI:10.1037/a0014726.

9. Baetjer, Howard, "Capital as Embodied Knowledge: Some Implications for the Theory of Economic Growth," *Review of Austrian Economics 13*, no. 2 (2000): 147–74.

10. Sherman, Lauren E., Leanna M. Hernandez, Patricia M. Greenfield, and Mirella Dapretto, "'What the Brain 'Likes': Neural Correlates of Providing Feedback on Social Media," *Social Cognitive and Affective Neuroscience 13*, no. 7 (July 2018): 699–707. DOI: 10.1093/scan/nsy051.

11. Edwards, Jonathan, "'13-year-old Boy Made and Trafficked Ghost Guns,' Authorities Say, and then Killed his Sister with One," *Washington Post* (December 3, 2021). https://www.washingtonpost.com/nation/2021/12/03/13-year-old-ghost-gun-trafficker/.

12. Pyne, Irene, "YouTubers Under 18 Who Are Earning up to US$22 Million per Year: Ryan Toysreview, Boram Tube Vlog and More," *Style*, October 17, 2019.

13. Alexander, Kristopher, interview by Loretta L.C. Brady, Zoom interview, Manchester, New Hampshire, November 2021.

14. Gray, Kishonna, *Intersectional Tech* (Louisiana: LSU Press, 2020), 105.

15. Chambers, Jennifer, "Oxford Officials Clarify Rumors about Nov. 30 School Shooting." *Detroit News* (January 19, 2022). https://www.detroitnews.com/story/news/local/oakland-county/2022/01/19/oxford-school-shooting-ethan-crumbley-social-media-posts-administrators-unaware/6578332001/.

16. Gray, Kishonna, *Intersectional Tech* (Louisiana: LSU Press, 2020).

17. Ibid.

18. Villegas, Seth, interview by Loretta L.C. Brady, Zoom interview, Manchester, New Hampshire, October 2021.

19. Wells, Georgia, and Jeff Horwitz, "The Facebook Files: Facebook's Effort to Attract Preteens Reaches Back Years—Documents Show Moves Came in Response to Competition from Snapchat, TikTok," *Wall Street Journal* (September 2021). https://www.proquest.com/docview/2577278884?accountid=13640; "Watch: House Hearing on Social Media Reform, with Facebook Whistleblower Frances Haugen," *PBS News Hour* (December 2021). https://www.pbs.org/newshour/economy/watch-live-house-hearing-on-social-media-reforms-with-facebook-whistleblower-frances-haugen.

20. "Adolescents 2030," *PMNCH for Women's, Children's, and Adolescent's Health.* https://www.adolescents2030.org/, accessed November 23, 2021.

CHAPTER 11

1. Berman, Robby, "The Worst Time of Middle Age? When You're 47," *BIG THINK* (January 2020). https://bigthink.com/neuropsych/middle-age-slump/.

2. Hollingworth, S., Ayo Mansaray, K. Allen, and A. Rose, "Parents' Perspectives on Technology and Children's Learning in the Home: Social Class and the Role of the Habitus," *Journal of Computer Assisted Learning* 27, no. 4 (August 2011): 347–60. DOI: 10.1111/j.1365-2729.2011. 00431.x.

3. Perry, Tam E., Luke Hassevoort, Nicole Ruggiano, and Natali Shtompel, "Applying Erikson's Wisdom to Self-Management Practices of Older Adults: Finding from Two Field Studies," *Research on Aging* 37, no. 3 (April 2015): 253–74. DOI: 10.1177/016402751427974.

4. Jay, Megan, *The Defining Decade: Why Your Twenties Matter and How to Make the Most of Them Now* (New York: Twelve, 2012).

5. Prilleltensky, Isaac, "Mattering at the Intersection of Psychology, Philosophy, and Politics," *American Journal of Community Psychology* 65, no. 1–2 (2019): 16–34. DOI: 10.1002/ajcp.12368.

6. Larissa Buhler, Janina, Rebekka Weidmann, Jana Nikitin, and Alexandra Grob, "A Closer Look at Life Goals Across Adulthood: Applying a Developmental Perspective to Content, Dynamics, and Outcomes of Goal Importance and Goal Attainability," *European Journal of Personality 33*, no. 3 (May 2019), 359–84. DOI: 10.1002/per.2194.

7. Berman, Robby, "The Worst Time of Middle Age? When You're 47," *BIG THINK* (January 2020). https://bigthink.com/neuropsych/middle-age -slump/.

8. Nelson, Garrison, "The Rise of the 'Straight, White, Angry Male' Voter," paper presented at Broken: Barriers, Parties, and Conventional Wisdom in 2016: American Elections Academic Symposium, Saint Anselm College, March 17–18, 2017. https://www.anselm.edu/sites/default/files/Documents /NHIOP/Symposium%20Program%202017.pdf.

9. Main, Thomas J., "Illiberalism in American Political Culture Today," from *The Rise of Illiberalism* (Washington, DC: Brookings Institution Press, 2021).

10. Pape, Robert A., and Keven Ruby. "The Face of American Insurrection: Right Wing Organizations Evolving into a Violent Mass Movement," Chicago Project on Security and Threats, The University of Chicago, January 28, 2021. https://d3qi0qp55mx5f5.cloudfront.net/cpost/i/docs/americas_insurrection ists_online_2021_01_29.pdf?mtime=1611966204.

11. Fulton III, Scott, "Is Social Media an Influential Technology or an Insurrectionist Tool? Status Report," *ZDNET* (January 2021). https://www.zdnet .com/article/is-social-media-an-influential-technology-or-an-insurrectionist -tool-status-report/.

12. Suderman, Alan, and Joshua Goodman, "Amid the Capitol Riot, Facebook Faced Its Own Insurrection," *AP NEWS* (October 2021). https:// apnews.com/article/donald-trump-technology-business-social-media-media -07124025bdbeba98a7c7b181562c3c1a.

13. Smith, Yves, "Michael Hudson: On Debt Parasites" [podcast], *Naked Capitalism* (January 2022). https://www.nakedcapitalism.com/2022/01/michael -hudson-on-debt-parasites.html.

14. Helson, Ravenna, and Laurel McCabe, "The Social Clock Project in Middle Age," from B.F. Turner and L.E. Troll (eds.), *Women Growing Older: Psychological Perspectives* (Thousand Oaks, CA: Sage Publications, 1994), 68–93.

15. Helson, Ravenna M., and Valory Mitchell, *Women on the River: A Fifty-Year Study of Adult Development* (Oakland: University of California Press, 2020), 237.

16. Alexander, Kristopher, interview by Loretta L.C. Brady, Zoom interview, Manchester, New Hampshire, November 2021.

17. Perry, Tam E., Luke Hassevoort, Nicole Ruggiano, and Natali Shtompel, "Applying Erikson's Wisdom to Self-Management Practices of Older Adults:

Finding from Two Field Studies," *Research on Aging 37*, no. 3 (April 2015): 253–74. DOI: 10.1177/016402751427974.

18. Nesse, Randolph M., "The Evolution of Hope and Despair," *Social Research 66*, no. 2 (Summer 1999): 429–69. http://www.jstor.org/stable /40971332.

19. Beardow, Jye. "Scroll, Click, Like, Share, Repeat: The Algorithmic Polarisation Phenomenon." *ANU Journal of Law and Technology 2*, no. 1 (2021): 153–164.

20. Mullaney, Thomas S., Benjamin Peters, Mar Hicks, and Kavita Philip, *Your Computer Is on Fire* (Cambridge, MA: The MIT Press, 2021).

21. Gray, Kishonna L., *Intersectional Tech: Black Users in Digital Gaming* (Louisiana: LSU Press, 2020), 157.

22. Ibid., 167.

23. Dreyfuss, Emily Robin, interview by Loretta L.C. Brady, Zoom interview, Manchester, New Hampshire, November 2021.

24. Ibid.

25. Gray, Kishonna L., *Intersectional Tech: Black Users in Digital Gaming* (Louisiana: LSU Press, 2020), 166.

26. The styles of writing Web 3.0 vary between Web 3.0, Web 3, or Web3. Ultimately, Web 3 (or Web 3.0) represents a decentralized internet grounded in tech including blockchain, cryptocurrency, and encryption, and has also been referred to as the semantic web.

CHAPTER 12

1. This image is most certainly composed of a combination of storylines and themes in the series *Mad Men*. It is a narrative cheat of one of the series closing scenes wherein Don Draper—the ad man depicted who saved a large tobacco account, contributing to the cancer that would take the men and women of his generation when wars alone did not—is seen at a 1970s commune, the green, love, and Jesus freak revolution visibly on the horizon of the Brylcreemed "corporate man."

2. Tsing, Anna, "Arts of Inclusion, or How to Love a Mushroom," *Mānoa 22*, no. 2, (Winter 2010): 191–203. https://www.jstor.org/stable/41479491.

3. Mitman, Gregg, "A Reflection of Plantationocene: A Conversation with Donna Haraway and Anna Tsing," *Edge Effects* (June 2019). https://edgeeffects .net/wp-content/uploads/2019/06/PlantationoceneReflections_Haraway _Tsing.pdf.

4. Istanbul Arastirmalari Enstitusu, "Anna L. Tsing—The Particular in the Planetary: Reimagining Cosmopolitanism Beyond the Human," YouTube video, 1:33:30, April 8, 2021. https://www.youtube.com/watch?v=PqLiEo3baMc.

5. Rooks, Belvie, Kabir Hypolite, Loretta Brady, and Bithiah Carter, "Panel 6, I'll Fly Away: How We Move Forward from Here. In Honor of Anya Dillard," virtual panel, Crossing River Jordan: Healing Racial Wounds Through Accountability and Truth Telling, October 23, 2021.

6. "Merrimack River at Risk," PBSNH Video, 56:14, July 14, 2020. https://www.pbs.org/video/the-merrimack-river-at-risk-cq7o6d/.

7. Chernow, Ron, *Alexander Hamilton* (New York: Penguin Books, 2005); Miranda, Lin-Manuel, "Hamilton: An American Musical," in *Hamilton: The Revolution*, edited by Jeremy McCarter (New York: Grand Central Publishing, 2016).

8. Hartmans, Avery, and Paige Leskin, "The Life and Rise of Travis Kalanick, Uber's Controversial Billionaire Co-Founder and CEO," *Insider* (May 2019). https://www.businessinsider.com/uber-ceo-travis-kalanick-life-rise-photos-2017-6#after-san-francisco-uber-rapidly-expanded-its-services-to-other-us-cities-in-may-2011-uber-launched-in-new-york-city-now-one-of-ubers-biggest-markets-more-than-168000-uber-rides-are-hailed-in-new-york-city-every-day-19.

9. Fincher, David, director, *The Social Network*, Columbia Pictures, 2010, 2:00:46. https://www.netflix.com/watch/70132721?trackId=13752289&tctx=0%2C0%2C08fefdc734e1fd2290fe0323f9162c02f590c1c7%3Abf3a2479d57bb9a0a89cccdfd3d7872e38b70ba1%2C08fefdc734e1fd2290fe0323f9162c02f590c1c7%3Abf3a2479d57bb9a0a89cccdfd3d7872e38b70ba1%2Cunknown%2C%2C%2C.

10. Newcomer, Eric, and Brad Stone, "The Fall of Travis Kalanick Was a Lot Weirder and Darker Than You Thought," *Bloomberg Businessweek* (January 2018). https://www.bloomberg.com/news/features/2018-01-18/the-fall-of-travis-kalanick-was-a-lot-weirder-and-darker-than-you-though.

11. Goldman, David, "Uber Strips Power from Ousted CEO Travis Kalanick," *CNN Business* (October 2017). https://money.cnn.com/2017/10/03/technology/business/uber-board-kalanick/index.html.

12. Uber Team, "Uber's New CEO," Uber Newsroom (August 2017). https://www.uber.com/newsroom/ubers-new-ceo-3/.

13. Fowley, Geoffrey A., "Uber CEO Q&A: When Rape Happens in an Uber, Who's Responsible?" *Washington Post* (December 6, 2019). https://www.washingtonpost.com/technology/2019/12/06/uber-ceo-qa-when-rape-happens-an-uber-whos-responsible/.

14. Parida, Tulsi, and Aparna Ashok, "Consolidating Power in the Name of Progress: Technosolutionism and Farmer Protests in India," from *Fake AI* (Manchester, UK: Meatspace Press, 2021), 164. https://ia801504.us.archive.org/6/items/fake-ai/Fake_AI.pdf.

15. Ullman, Ellen, *Life in Code: A Personal History of Technology* (New York: MCD, 2017); Ullman, Ellen, *Close to the Machine: Technophilia and Its Discontents* (New York: Picadorusa, 2012).

16. Greenfield, Patricia M. "Linking Social Change and Developmental Change: Shifting Pathways of Human Development." *Developmental Psychology* 45, no. 2 (2009): 401–18. DOI:10.1037/a0014726.

17. Gould was fourteen when he left his family home to make his way in the world; only four years earlier, he had survived the "renters war," when tenants of his family land committed violence and refused to pay their rent.

18. The story of the Erie Wars includes a million-dollar swindle involving kidnapping, foreign invasion, suicide, and a canal and Senate seat.

19. Markus, Hazel Rose, and Alana Conner, *Clash!: How to Thrive in a Multicultural World* (New York: Penguin Group, 2014).

20. Mueller, Robert S., III. "Report on the Investigation into Russian Interference in the 2016 Presidential Election," US Department of Justice, Vol. I (March 2019). https://www.justice.gov/archives/sco/file/1373816/download.

21. Hampton, Rachelle, "The Black Feminists Who Saw the Alt-Right Threat Coming," *Slate*, April 2019. https://slate.com/technology/2019/04/black-feminists-alt-right-twitter-gamergate.html.

22. Parida, Tulsi, and Aparna Ashok, "Consolidating Power in the Name of Progress: Technosolutionism and Farmer Protests in India," from *Fake AI* (Manchester, UK: Meatspace Press, 2021), 165. https://ia801504.us.archive.org/6/items/fake-ai/Fake_AI.pdf.

23. Knight, Matthew, "'Miracle Material' Chips Away at Silicon Dominance," *CNN Business* (October 2011). https://www.cnn.com/2011/10/12/tech/graphene-computer-chip-research/index.html.

24. Patton, Desmond, interview by Loretta L.C. Brady, Zoom interview, Manchester, New Hampshire, December 2021.

25. Ibid.

26. Bronfenbrenner, Urie. "Developmental Research, Public Policy, and the Ecology of Childhood." *Child Development* 45, no. 1 (1974): 1–5.

CHAPTER 13

1. "Urie Bronfenbrenner," College of Human Ecology. https://bctr.cornell.edu/about-us/urie-bronfenbrenner.

2. Kamp, David, *Sunny Days: The Children's Television Revolution That Changed America* (New York: Simon & Schuster, 2020).

3. Langlois, Richard N. "'An Elephants' Graveyard': The Deregulation of American Industry in the Late Twentieth Century," University of Connecticut (July 2021). https://media.economics.uconn.edu/working/2021-11.pdf.

4. Rothstein, Richard, *The Color of Law: A Forgotten History of How Our Government Segregated* (New York: Liveright Publishing, 2017).

5. Horwitz, Robert Britt, "Telecommunications and Their Deregulation: An Introduction," from *The Irony of Regulatory Reform: The Deregulation of American Telecommunication* (New York: Oxford University Press, 1989), 1–44

6. Poole, Mary Elizabeth, "Securing Race and Ensuring Dependence: The Social Security Act of 1935" (Rutgers, The State University of New Jersey— New Brunswick, 2000).

7. Avery, James M., and Mark Peffley, "Race Matters: The Impact of News Coverage of Welfare Reform on Public Opinion," *Race and the Politics of Welfare Reform* (2003): 13–50.

8. Horwitz, Robert Britt, *The Irony of Regulatory Reform: The Deregulation of American Telecommunication* (New York: Oxford University Press, 1989).

9. Christensen, Clayton M., and Michael B. Horn, "How Do We Transform Our Schools? Use Technologies that Compete Against Nothing," *Education Next* 8, no. 3 (2008): 12–20.

10. Horwitz, Robert Britt, *The Irony of Regulatory Reform: The Deregulation of American Telecommunication* (New York: Oxford University Press, 1989).

11. Hetzel, Robert L., "Launch of the Bretton Woods System," *Federal Reserve History* (2013).

12. McCarthy, John, Marvin L. Minsky, Nathaniel Rochester, and Claude E. Shannon. "A Proposal for the Dartmouth Summer Research Project on Artificial Intelligence, August 31, 1955," *AI* magazine 27, no. 4 (2006): 12.

13. Bonvillian, William B., "The Rise of Advanced Manufacturing Institutes in the United States," *The Next Production Revolution: Implications for Governments and Business* (2017): 55.

14. Donohue, Chip, interview by Loretta L.C. Brady, Zoom interview, Manchester, New Hampshire, October 2021.

15. UN News Staff, "Urgent Action Needed Over Artificial Intelligence Risks to Human Rights," *UN News*. September 21, 2021. https://news.un.org /en/story/2021/09/1099972.

16. Borokini, Favour, and Ridwan Oloyede, "When Fintech Meets Sixty Million Unbanked Citizens," from *Fake AI* (Poland: Meatspace Press, 2021), 170.

17. Donohue, Chip, interview by Loretta L.C. Brady, Zoom interview, Manchester, New Hampshire, October 2021.

18. Dreyfuss, Emily Robin, interview by Loretta L.C. Brady, Zoom interview, New Hampshire, November 2021.

19. "The Cloud Is a Factory," from *Your Computer Is on Fire*, ed. Thomas S. Mullaney, Benjamin Peters, Mar Hicks, and Kavita Philip (Cambridge, MA: The MIT Press, 2021): 29–50; Harvey, Adam, "What Is a Face," from *Fake AI* (Manchester, UK: Meatspace Press, 2021), 135–44.

20. Dreyfuss, Emily Robin, interview by Loretta L.C. Brady, Zoom interview, Manchester, New Hampshire, November 2021.

21. UN News Staff, "Urgent Action Needed Over Artificial Intelligence Risks to Human Rights," *UN News*, September 21, 2021. https://news.un.org/en/story/2021/09/1099972.

22. Held, Virginia, *The Ethics of Care: Personal, Political, and Global* (Oxford: Oxford University Press, 2006).

CHAPTER 14

1. Fraser, Jennifer, and Anton Yasnitsky, "Deconstructing Vygotsky's Victimization Narrative: A Re-examination of the 'Stalinist Suppression' of Vygotskian Theory," *History of the Human Sciences* 28, no. 2 (2015): 128–53. http://individual.utoronto.ca/yasnitsky/texts/Fraser%20%26%20Yasnitsky%20(2015).pdf.

2. "Make Way for Ducklings," *International Travel News*. https://www.intltravelnews.com/2010/09/go-east-young-duckshttps://www.intltravelnews.com/2010/09/go-east-young-ducks.

3. Lewis, Danny, "Reagan and Gorbachev Agreed to Pause the Cold War in Case of an Alien Invasion," *Smithsonian* (November 2015). https://www.smithsonianmag.com/smart-news/reagan-and-gorbachev-agreed-pause-cold-war-case-alien-invasion-180957402/.

4. Vygotsky, Lev S., "Play and its Role in the Mental Development of the Child," *Soviet Psychology* 5, no. 3 (1967): 6–18.

5. Crockett, Lisa J., "Cultural, Historical, and Subcultural Context of Adolescence: Implications of Health and Development," *Faculty Publications, Department of Psychology University of Nebraska* (June 1997), 23–53. https://digitalcommons.unl.edu/cgi/viewcontent.cgi?article=1243&context=psychfacpub.

6. Oduro, Serena Dokuaa, "Do We Need AI or Do We Need Black Feminism? A Poetic Guide," from *Fake AI* (Manchester, UK: Meatspace Press, 2021), 89–95. https://ia801504.us.archive.org/6/items/fake-ai/Fake_AI.pdf.

7. Straight, Andrew, "Why Automated Content Moderation Won't Save Us," from *Fake AI* (Manchester, UK: Meatspace Press, 2021), 148–58. https://ia801504.us.archive.org/6/items/fake-ai/Fake_AI.pdf; 148.

8. Ibid., 158.

9. Ibid., 95; Gorwa, Robert, Reuben Binns, and Christian Katzenbach, "Algorithmic Content Moderation: Technical and Political Challenges in the Automation of Platform Governance," *Big Data & Society* 7, no. 1 (2020): 1–15. DOI: 10.1177/2053951719897945.

10. Thakor, Mitali, "Capture Is Pleasure," from *Your Computer Is on Fire*, ed. Thomas S. Mullaney, Benjamin Peters, Mar Hicks, and Kavita Philip (Cambridge, MA: The MIT Press, 2021).

11. Ibid.

12. Ensenmenger, Nathan, "The Cloud Is a Factory," from *Your Computer Is on Fire*, ed. Thomas S. Mullaney, Benjamin Peters, Mar Hicks, and Kavita Philip (Cambridge, MA: The MIT Press, 2021), 29–50.

13. The proposed project would have been vetted by an Institutional Review Board had funding been granted, and all our approaches to using the iBeacons, incentivizing community visits in locations without substances, and limitations to investigators and app developers on user data access were actively negotiated as we planned the project. Still, the idea that local recovering people were being tracked in their local communities was not something I considered lightly, no matter how compelling the power of such reinforcements for positive social behavior might have been from a treatment perspective. The ethics of undertreated opioid addiction was also a factor in considering the project. Regardless, it was not funded, and nothing further occurred. Within a year the company with the app had closed its doors, moving on to other ventures.

14. Kunzru, Hari, "You Are Cyborg," *Wired* (February 1997). https://www.wired.com/1997/02/ffharaway/.

15. Space, as in both celestial and influential, each assisting his monopolistic aims in synchrony.

16. Matz, S.C., M. Kosinski, G. Nave, and D.J. Stillwell, "Psychological Targeting as an Effective Approach to Digital Mass Persuasion," *PNAS 114*, no. 48 (Nov. 2017), 12714–19. DOI: 10.1073/pnas/170966114; Sudermann, Alan, and Joshua Goodman, "Amid the Capitol Riot, Facebook Faced Its Own Insurrection," *AP News* (October 2021). https://apnews.com/article/donald-trump-technology-business-social-media-media-07124025bdbe ba98a7c7b181562c3c1a.

17. Edwards, Paul N., "Platforms Are Infrastructures on Fire," *Your Computer Is on Fire*, ed. Thomas S. Mullaney, Benjamin Peters, Mar Hicks, and Kavita Philip (Cambridge, MA: The MIT Press, 2021).

18. Dreyfuss, Emily Robin, interview by Loretta L.C. Brady, Zoom interview, Manchester, New Hampshire, November 2021.

19. Leonardo, Zeus, and Logan Manning, "White Historical Activity Theory: Toward a Critical Understanding of White Zones of Proximal Development," *Race and Ethnicity and Education* (November 2015). DOI: 10.1080/13613324.2015.1100988.

20. Patton, Desmond, interview by Loretta L.C. Brady, Zoom interview, Manchester, New Hampshire, December 2021.

21. Alexander, Kristopher, interview by Loretta L.C. Brady, Zoom interview, Manchester, New Hampshire, November 2021.

CHAPTER 15

1. Colapinto, John, "Material Questions," *New Yorker* (December 2014). https://www.newyorker.com/magazine/2014/12/22/material-question.

2. Callahan, Daniel, "Bioethics and Policy—A history," Hastings Center, Brief Report series. https://www.thehastingscenter.org/briefingbook/bioethics -and-policy-a-history/, published September 15, 2015.

3. Ibid.

4. Patton, Desmond, interview, conducted by Loretta L.C. Brady, December 2021.

5. Donohue, Chip, interview by Loretta L.C. Brady, Zoom interview, Manchester, New Hampshire, October 2021.

6. "A Federal Strategy to Ensure Secure and Reliable Supplies of Critical Minerals," US Department of Commerce (June 2019). https://www.commerce.gov /sites/default/files/2020-01/Critical_Minerals_Strategy_Final.pdf.

7. How COVID-19 changed that reality!

8. Leonardo, Zeus, and Logan Manning, "White Historical Activity Theory: Toward a Critical Understanding of White Zones of Proximal Development," *Race and Ethnicity and Education* (November 2015). DOI: 10.1080/13613324.2015.1100988.

9. "As women, we have been taught either to ignore our differences, or to view them as causes for separation and suspicion rather than as forces for change. Without community, there is no liberation, only the most vulnerable and temporary armistice between an individual and her oppression. But community must not mean a shedding of our differences, nor the pathetic pretense that these differences do not exist." Lorde, Audre. "The Master's Tools Will Never Dismantle the Master's House," from *Sister Outsider: Essays and Speeches* (Berkeley, CA: Crossing Press, 1984), 111.

10. Wells, Georgia, Jeff Horwitz, and Deepa Seetharaman, "The Facebook Files: Facebook Knows Instagram Is Toxic for Teen Girls, Its Research Shows— Internal Document Shows a Youth Mental-Health Issue that Facebook Plays Down in Public," *Wall Street Journal* (September 2021). https://www.proquest .com/docview/2572410901?accountid=13640&parentSessionId=YW4qkaAdJ brpLHiL5TY20t%2BAJKN7xytSr0ayv70%2BKMg%3D; and Wells, Georgia, and Jeff Horwitz, "The Facebook Files: Facebook's Effort to Attract Preteens Reaches Back Years—Documents Shows Moves Came in Response to Competition from Snapchat, TikTok," *Wall Street Journal* (September 2021). https:// www.proquest.com/docview/2577278884?accountid=13640; and Scheck, Justin, Newley Purnell, and Jeff Horwitz, "The Facebook Files: Facebook's Staff Flags Criminals, But Company Often Fails to Act—Documents Show Alarm About What Is on the Site Outside U.S., Where User Numbers Are Huge,"

Wall Street Journal (September 2021). https://www.proquest.com/docview/257
3296131?accountid=13640; and Horwitz, Jeff, "The Facebook Files: Facebook
Documents Reveal Secret Elite Exempt from Its Rules—XCheck Program
Gives Celebrities, Politicians Special Treatment, Which Some Abuse," *Wall
Street Journal* (September 2021). https://www.proquest.com/docview/257211
6186?accountid=13640.

11. Collins, Patricia Hill, "The Difference That Power Makes: Intersectionality and Participatory Democracy," *Investigaciones Feministas 8*, no. 1 (February
2017): 19–39. DOI: 10.5209/INFE.54888.

12. Oduro, Serena, "A New AI Lexicon: Black Women Best," *Medium*
(September 2021). https://medium.com/a-new-ai-lexicon/a-new-ai-lexicon
-black-women-best-14eb5d059b2a.

13. Oduro, Serena Dokuaa, "Do We Need AI or Do We Need Black
Feminism? A Poetic Guide," from *Fake AI* (Manchester, UK: Meatspace Press,
2021), 89–95. https://ia801504.us.archive.org/6/items/fake-ai/Fake_AI.pdf.

Bibliography

"About," McNair Scholars McNair Scholars Program. https://mcnairscholars.com/about/.

Adam, Harvey. "What Is a Face," from *Fake AI* (Poland: Meatspace Press, 2021), 135–44.

Addams, Jane. "The Devil Baby at Hull House," from *The Best American Essays of the Century*. Oates, Joyce Carol, and Robert Atwan, editors (Boston: Houghton Mifflin Company, 2001), 75–89.

"Adolescents 2030." *PMNCH for Women's, Children's, and Adolescent's Health.* https://www.adolescents2030.org/, accessed November 23, 2021.

"A Federal Strategy to Ensure Secure and Reliable Supplies of Critical Minerals." US Department of Commerce (June 2019). https://www.commerce.gov/sites/default/files/2020-01/Critical_Minerals_Strategy_Final.pdf.

Ainsworth, Mary S., and John Bowlby. "An Ethological Approach to Personality Development." *American Psychologist 46*, no. 4 (1991): 333.

Alexander, Kristopher. Interview by Loretta L.C. Brady. Zoom interview. Manchester, New Hampshire, November 2021.

Almond, Paul, director. *Seven Up.* Granada Television. 1964 (London, UK).

Armheim, Rudolf. *Visual Thinking* (Berkeley: University of California Press, 1969).

"Assessment of Adult Attachment," from Daniel P. Brown and David S. Elliott, editors, *Attachment Disturbances in Adults: Treatment for Comprehensive Repair* (New York: W. W. Norton, 2016).

Auxier, Brooke, Monica Anderson, Andrew Perrin, and Erica Turner. "Parenting Children in the Age of Screens." Pew Research Center (July 2020). https://www.pewresearch.org/internet/2020/07/28/parenting-children-in-the-age-of-screens/.

Avery, James M., and Mark Peffley. "Race Matters: The Impact of News Coverage of Welfare Reform on Public Opinion." *Race and the Politics of Welfare Reform* (2003): 131–50.

Baccarella, Christian V., Timm F. Wagner, Jan F. Kietzmann, and Ian P. Mc-Carthy. "Ascertaining the Rise of the Dark Side of Social Media: The Role of Sensitization and Regulation." *European Management Journal 38*, no. 1 (February 2020): 3–6. DOI: 10.1016/j.emj.2019.12.011.

Baetjer, Howard. "Capital as Embodied Knowledge: Some Implications for the Theory of Economic Growth." *Review of Austrian Economics 13*, no. 2 (2000): 147–74.

Baird, A.A. "Understanding Adolescent Decision Making" (invited talk). Association for Psychological Science, Annual Meeting, May 2007, Chicago, Illinois.

Bandura, Albert, Joan E. Grusec, and Frances L. Menlove. "Observational Learning as a Function of Symbolization and Incentive Set." *Child Development* (1966): 499–506.

Barnett, Clive. "Technologies of Citizenship: Assembling Media Publics," from *Culture and Democracy: Media, Space, and Representation* (Edinburgh: Edinburgh University Press, 2003), 81–107.

Bedingfield, Will. "The Far Right Are Running Riot on Steam and Discord." *Wired.* December 2021. https://www.wired.co.uk/article/steam-discord-far-right?utm_source=twitter&utm_medium=social&utm_campaign=onsite-share&utm_brand=wired-uk&utm_social-type=earned.

Belkin, Lisa. "The Newest, Latest Parenting Trend." *Motherlode Blog.* November 8, 2010. https://parenting.blogs.nytimes.com/2010/11/08/the-newest-latest-parenting-trend/.

Berkowitz, Talia, Marjorie W. Schaeffer, Erin A. Maloney, Lori Peterson, Courtney Gregor, Susan C. Levine, and Sian L. Beilock. "Math at Home Adds Up to Achievement in School." *Science,* 350, no. 6257 (October 2015): 196–8. DOI:10.1126/science.aac7427.

Berman, Robby. "The Worst Time of Middle Age? When You're 47." *BIG THINK* (January 2020). https://bigthink.com/neuropsych/middle-age-slump/.

Bonvillian, William B. "The Rise of Advanced Manufacturing Institutes in the United States." *The Next Production Revolution: Implications for Governments and Business* (2017): 55.

Borokini, Favour, and Ridwan Oloyede. "When Fintech Meets 60 Million Unbanked Citizens," from *Fake AI* (Manchester, UK: Meatspace Press, 2021), 170.

Brady, Loretta L.C. "Canaries in the Ethical Coal Mine? Case Vignettes and Empirical Findings for How Psychology Leaders Have Adopted Twitter." *Ethics and Behavior 26*, no. 2 (February 2016): 110–27. DOI: 10.1080/10508422.2014.994064.

Brady, Loretta, and Elizabeth Evans. "Financial Market Shifts in the Era of Regulation, Deregulation, and Global Markets: How an Ethic of Care Lens Can Reveal and Mitigate Financial Market Risks and the Ethical Responsibility This Entails." Presented at the first annual *The Ethics of Business, Trade,*

& Global Governance: An Interdisciplinary Conference, December 1, 2018, New Castle, New Hampshire.

Brady, L.L.C., and E. P. Ossoff. "Fifty Years of Fashion and Feminism: The Effect of Early Recognition on Vocational and Sociopolitical Identity Development." *Current Psychology 29*, no. 10 (2010): 34–44.

Brady, Loretta L.C., Rebecca Hadley, and Cathy Kuhn. "Creating a Family-Centered Wellness Team: Lessons Learned in Creating and Integrated Continuum of Care for Families Facing Homelessness, Addiction, and Trauma Recovery." *Journal of Social Distress and Homelessness 19*, no. 1–2 (July 2013): 83–106. DOI: 10.1179/105307809805365163.

Brashier, N. M., and Schacter, D. L. "Aging in an Era of Fake News." *Current Directions in Psychological Science 29*, no. 3 (2020): 316-23. DOI: 10.1177/0963721420915872.

"Brazelton Touchpoint Center." Brazelton Touchpoint Center. https://www.brazeltontouchpoints.org/.

Brazelton, T. Berry, and Joshua D. Sparrow. *Touchpoints—Birth to Three: Your Child's Emotional and Behavioral Development*, 2nd edition (Cambridge, MA: Da Capo Lifelong Books, 2006), 373, 344, 371, 446.

Brazelton, T. Berry, and Joshua D. Sparrow. *Touchpoints—Three to Six: Your Child's Emotional and Behavioral Development* (Cambridge, MA: Da Capo Press, 2001), 341–371.

Brodie, Richard. *Virus of the Mind: The New Science of the Meme* (Carlsbad, CA: Hay House, Inc., 2011).

Browne, Angela, Brenda Miller, and Eugene Maguin. "Prevalence and Severity of Lifetime Physical and Sexual Victimization Among Incarcerated Women." *International Journal of Law and Psychiatry 22*, no. 3–4 (1999): 301–22..

Bruner, Jerome. *Acts of Meaning* (Boston: Harvard University Press, 1990).

Callahan, Daniel. "Bioethics and Policy—A History." Hastings Center. Brief Report series, September 15, 2015. https://www.thehastingscenter.org/briefingbook/bioethics-and-policy-a-history/.

Calvert, Sandra L., and Melissa N. Richards. "Children's Parasocial Relationships," from A. Jordon and D. Romer, editors, *Media and the Well-Being of Children and Adolescents* (New York: Oxford University Press, 2014), 187–200.

Chambers, Jennifer. "Oxford Officials Clarify Rumors about Nov. 30 School Shooting." *Detroit News*, January 19, 2022. htpps://www.detroitnews.com/story/news/local/oakland-county/2022/01/19/oxford-school-shooting-ethan-crumbley-social-media-posts-administrators-unaware/6578332001/.

Chernow, Ron. *Alexander Hamilton* (New York: Penguin Books, 2005).

Christensen, Brandon. "10 Worst Space Disasters in History." *Real Clear History*. March 2019. https://www.realclearhistory.com/articles/2019/03/08/10_worst_space_disasters_in_history_420.html.

Christensen, Clayton M., and Michael B. Horn. "How Do We Transform Our Schools? Use Technologies That Compete Against Nothing." *Education Next* 8, no. 3 (2008): 12–20.

Choney, Suzanne. "Apple Offers iTunes Credits to Parents for In-App Purchases Made by Kids." June 24, 2013. https://www.nbcnews.com/tech/mobile /apple-offers-itunes-credits-parents-app-purchases-made-kids-f6C10424015.

Colapinto, John. "Material Questions." *New Yorker* (December 2014). https:// www.newyorker.com/magazine/2014/12/22/material-question.

Collins, Patricia Hill. *Another Kind of Public Education: Race, Schools, the Media, and Democratic Possibilities* (Boston: Beacon Press, 2009).

Collins, Patricia Hill. "The Difference That Power Makes: Intersectionality and Participatory Democracy." *Investigaciones Feministas 8*, no. 1 (February 2017): 19–39. DOI: 10.5209/INFE.54888.

Confessore, Nicholas. "Cambridge Analytica and Facebook: The Scandal and the Fallout So Far." *New York Times*. April 4, 2018. https://www.nytimes .com/2018/04/04/us/politics/cambridge-analytica-scandal-fallout.html.

Craig, Maureen A., and Richeson, Jennifer A. "Information about the US Racial Demographic Shift Triggers Concerns about Anti-White Discrimination Among the Prospective White 'Minority.'" *PloS One 12*, no. 9 (2017). DOI: e0185389.

Crockett, Lisa J. "Cultural, Historical, and Subcultural Context of Adolescence: Implications of Health and Development." *Faculty Publications, Department of Psychology University of Nebraska* (June 1997): 23–53. https://digitalcommons .unl.edu/cgi/viewcontent.cgi?article=1243&context=psychfacpub.

Cromer, Jason A., Adrian J. Schembri, Brian T. Harel, and Paul Maruff. "The Nature and Rate of Cognitive Maturation from Late Childhood to Adulthood." *Frontiers in Psychology* (May 2015). DOI: 10.3389/fpsyg.2015.00704.

Cumbers, John. "The Social Dilemma: What Can the Biotech Industry Learn from the Failures of the Social Media Industry?" *Forbes* (October 2020). https://www.forbes.com/sites/johncumbers/2020/10/15/the-social -dilemma-what-can-the-biotech-industry-learn-from-the-failures-of-the -social-media-industry/?sh=c86caa752433.

Curtiss, Susan, Victoria Fronkin, Stephen Krashen, David Rigler, and Marilyn Rigler. "The Linguistic Development of Genie." *Language 50*, no. 3 (September 1974): 528–54. DOI: 10.2307/412222.

Darwin, Charles. "A Biographical Sketch of an Infant." *Mind 2*, no. 7 (1877): 285–94.

Darwin, Charles. *On the Origin of Species by Means of Natural Selection, or Preservation of Favored Races in the Struggle for Life* (London: John Murray, 1859).

Dawkins, Richard. "The Selfish Meme." *Time 153*, no. 15 (1999): 52–3.

DeFord, Carolyn. "Our Bodies Are Just a Shell: A Mother's Wisdom on Life And Death." *Story Corps*. https://storycorps.org/stories/our-bodies-are-just -a-shell-a-mothers-wisdom-on-life-and-death/.

DelGuidice, Marco. "Middle Childhood: An Evolutionary-Developmental Synthesis." *Handbook of Life Course Health Developmental*. ed. N. Halfon, C. Forrest, R. Lerner, and E. Fraustman (November 2017), 95–107. DOI: 10.1007/978-3-319-47143-3_5.

Dickinson, E.J. "What Is the Momo Challenge? Why Parents Are Freaking Out About This Terrifying 'Game.'" *Rolling Stone*. February 2019. https://www.rollingstone.com/culture/culture-news/what-is-momo-challenge-800470/.

Donohue, Chip, and Roberta Schomburg. "Technology and Interactive Media in Early Childhood Programs: What We've Learned from Five Years of Research, Policy, and Practice." *Young Children* 72, no. 4 (2017): 72–78.

Donohue, Chip. Interview by Loretta L.C. Brady. Zoom interview. Manchester, New Hampshire. October 2021.

Dreyfuss, Emily Robin. Interview by Loretta L.C. Brady. Zoom interview. Manchester, New Hampshire. November 2021.

Dugan, Therese E., Reed Stevens, and Siri Mehus. "From Show, to Room, to World: A Cross-Context Investigation of How Children Learn from Media Programming." *International Society of the Learning Sciences 1* (June 2010): 992–9. DOI: 10.22318/icls2010.1.992.

"Dust or Magic, Fall Institute via Zoom." Dust or Magic. http://dustormagic.com/institute/.

Edwards, Jonathan. "'13-year-old Boy Made and Trafficked Ghost Guns,' Authorities Say, and Then Killed His Sister with One." *Washington Post*. December 3, 2021. https://www.washingtonpost.com/nation/2021/12/03/13-year-old-ghost-gun-trafficker/.

Edwards, Paul N. "Platform Are Infrastructures on Fire." *Your Computer Is on Fire*. Thomas S. Mullaney, Benjamin Peters, Mar Hicks, and Kavita Philip, editors (Cambridge, MA: The MIT Press, 2021).

Ensenmenger, Nathan. "The Cloud Is a Factory," from *Your Computer Is on Fire*. Ed. Thomas S. Mullaney, Benjamin Peters, Mar Hicks, and Kavita Philip, editors. (Cambridge, MA: The MIT Press, 2021), 29–50.

Ferguson, Amber, and Kyle Sweson. "Texas Family Says Teen Killed Himself in Macabre 'Blue Whale' Online Challenge That's Alarming Schools." *Washington Post* (July 2017). https://www.washingtonpost.com/news/morning-mix/wp/2017/07/11/texas-family-says-teen-killed-himself-in-macabre-blue-whale-online-challenge-thats-alarming-schools/.

Fincher, David, director. *The Social Network*. Columbia Pictures. 2010. 2:00:46. https://www.netflix.com/watch/70132721?trackId=13752289&tctx=0%2C0%2C08fefdc734e1fd2290fe0323f9162c02f590c1c7%3Abf3a2479d57bb9a0a89cccdfd3d7872e38b70ba1%2C08fefdc734e1fd2290fe0323f9162c02f590c1c7%3Abf3a2479d57bb9a0a89cccdfd3d7872e38b70ba1%2Cunknown%2C%2C%2C.

Fitts, Alex Sobel. "Techs' Harassment Crisis Now an Arsenal of Smoking Guns." Backchannel. March 24, 2017. https://www.wired.com/2017/03/techs-harassment-crisis-now-has-an-arsenal-of-smoking-guns/.

Fowley, Geoffrey A. "Uber CEO Q&A: When Rape Happens in an Uber, Who's Responsible?" *Washington Post* (December 6, 2019). https://www.wash ingtonpost.com/technology/2019/12/06/uber-ceo-qa-when-rape-happens -an-uber-whos-responsible/.

Fraser, Jennifer, and Anton Yasnitsky. "Deconstructing Vygotsky's Victimization Narrative: A Re-examination of the 'Stalinist Suppression' of Vygotskian Theory." *History of the Human Sciences 28*, no. 2 (2015): 128–53. http:// individual.utoronto.ca/yasnitsky/texts/Fraser%20%26%20Yasnitsky%20 (2015).pdf.

Fry, Hannah. *Hello World: How to Be Human in the Age of the Machine* (New York: W.W. Norton & Company, 2019).

Fulton III, Scott. "Is Social Media an Influential Technology or an Insurrectionist Tool? Status Report." *ZDNET* (January 2021). https://www.zdnet.com /article/is-social-media-an-influential-technology-or-an-insurrectionist -tool-status-report/.

Gal, Noam, Limor Shifman, and Zohar Kampf. "'It Gets Better': Internet Memes and the Construction of Collective Identity." *New Media & Society* 18, no. 8 (2016): 1698–714.

Ghansah, Rachel Kaadzi. "A Most American Terrorist: The Making of Dylann Roof." *GQ.* August 2017. https://www.gq.com/story/dylann-roof-making -of-an-american-terrorist.

Gladwell, Malcolm. *The Tipping Point: How Little Things Can Make a Big Difference* (Boston: Little, Brown, 2006).

Goldman, David. "Uber Strips Power from Ousted CEO Travis Kalanick." *CNN Business.* (October 2017). https://money.cnn.com/2017/10/03/tech nology/business/uber-board-kalanick/index.html.

Gorwa, Robert, Reuben Binns, and Christian Katzenbach. "Algorithmic Content Moderation: Technical and Political Challenges in the Automation of Platform Governance." *Big Data & Society* (February 2020): 1–15. DOI: 10.1177/2053951719897945.

Graham, Ruth. "Baby's First Photo: The Unstoppable Rise of the Ultrasound Souvenir Industry." *Buzzfeed News* (September 2014). https://www.buzzfeed news.com/article/ruthgraham/sharing-ultrasound-photos-facebook-instagram.

Gray, Kishonna. *Intersectional Tech: Black Users in Digital Gaming.* (LSU Press, Louisiana, 2020), 105, 157, 166–7.

Greenfield, Patricia M. "Linking Social Change and Developmental Change: Shifting Pathways of Human Development." *Developmental Psychology 45*, no. 2 (2009): 401–18. DOI:10.1037/a0014726.

Guernsey, Lisa. Interview by Loretta L.C. Brady. Zoom interview. Manchester, New Hampshire. October 2021.

Guernsey, Lisa. *Screen Time: How Electronic Media—From Baby Videos to Educational Software—Affects Your Young Child* (New York: Basic Books, 2012).

Guz, Samantha. "The Stories We Tell Ourselves, Part 4: School Social Work and White Womanhood." *SSWN*. November 2, 2021. https://school socialwork.net/the-stories-we-tell-ourselves-part-4-school-social-work-and -white-womanhood/.

Halfon, Neal, and Miles Hochestein. "Life Course Health Development: An Integrated Framework for Developing Health, Policy, and Research." *Mil-Back Quarterly 80*, no. 3 (September 2002): 433–79. DOI: 10.1111/1468-0009.00019.

Hampton, Rachelle. "The Black Feminists Who Saw the Alt-Right Threat Coming." *Slate*. April 2019. https://slate.com/technology/2019/04/black -feminists-alt-right-twitter-gamergate.html.

Hartmans, Avery, and Paige Leskin. "The Life and Rise of Travis Kalanick, Uber's Controversial Billionaire Co-Founder and CEO." *Insider* (May 2019). https://www.businessinsider.com/uber-ceo-travis-kalanick-life-rise-photos -2017-6#after-san-francisco-uber-rapidly-expanded-its-services-to-other -us-cities-in-may-2011-uber-launched-in-new-york-city-now-one-of -ubers-biggest-markets-more-than-168000-uber-rides-are-hailed-in-new -york-city-every-day-19.

Haughton, Ciaran, Mary Aijen, and Carly Cheevers. "Cyber Babies: The Impact of Emerging Technology on the Developing Infant." *Psychology Research 5*, no. 9 (September 2015): 504-18. DOI: 10.17265/2159-5542/2015.09.002.

Hearing Before the United States Senate Committee on the Judicary and the United State Senate Committee on Commerce, Science, and Transportation. (2018). Statement of Mark Zuckerberg, chairman and chief executive officer of Facebook. https://www.judiciary.senate.gov/imo/media/doc/04 -10-18%20Zuckerberg%20Testimony.pdf.

Held, Virginia. *The Ethics of Care: Personal, Political, and Global* (Oxford: Oxford University Press, 2006).

Helson, Ravenna, and Laurel McCabe. "The Social Clock Project in Middle Age," in B.F. Turner and L.E. Troll, eds. *Women Growing Older: Psychological Perspectives* (Thousand Oaks, California: Sage Publications, 1994), 68–93.

Helson, Ravenna M., and Valory Mitchell. *Women on the River: A Fifty-Year Study of Adult Development* (Oakland: University of California Press, 2020), 237.

Hern, Alex. "Cambridge Analytica: How Did It Turn Clicks into Votes?" *Guardian*. May 6, 2018. https://www.theguardian.com/news/2018/may/06 /cambridge-analytica-how-turn-clicks-into-votes-christopher-wylie.

Hernandez, Javier C., and Albee Zhang. "90 Minutes in a Day, Until 10 P.M.: China Sets Rules for Young Gamers." *New York Times* (November 6, 2019). https://www.nytimes.com/2019/11/06/business/china-video-game -ban-young.html.

Hess, Eckhard Heinrich. *Imprinting: Early Experience and the Developmental Psychobiology of Attachment*, Foreword by Konrad Lorenz (New York: Van Nostrand Reinhold, 1973).

Hetzel, Robert L. "Launch of the Bretton Woods System." *Federal Reserve History* (2013).

Hirsh-Pasek, Kathy, Jennifer M. Zosh, Roberta Michnick Golinkoff, James H. Gray, Michael B. Robb, and Jordy Kaufman. "Putting Education in 'Educational' Apps: Lessons from the Science of Learning." *Association for Psychological Science 16*, no. 1 (April 2015): 3–34. DOI: 10.1177/1529100615569721.

History.com editors. "Space Shuttle *Challenger* Disaster." *History*. November 2009. https://www.history.com/this-day-in-history/challenger-explodes.

Hollingworth, S., Ayo Mansaray, K. Allen, and A. Rose. "Parents' Perspectives on Technology and Children's Learning in the Home: Social Class and the Role of the Habitus." *Journal of Computer Assisted Learning 27*, no. 4 (August 2011): 347–60. DOI: 10.1111/j.1365-2729.2011. 00431.x.

Hong, Yoo Rha, and Jae Sun Park. "Impact of Attachment, Temperament, and Parenting on Human Development." *Korean Journal of Pediatrics 55*, no. 12 (December 2012), 449–54. DOI:10.3345/kjp.2012.55.12.449.

Horwitz, Jeff. "The Facebook Files: Facebook Documents Reveal Secret Elite Exempt from Its Rules—XCheck Program Gives Celebrities, Politicians Special Treatment, Which Some Abuse." *Wall Street Journal* (September 2021). https://www.proquest.com/docview/2572116186?accountid=13640.

Horwitz, Robert Britt. *The Irony of Regulatory Reform: The Deregulation of American Telecommunication* (New York: Oxford University Press, 1989).

Horwitz, Robert Britt. "Telecommunications and Their Deregulation: An Introduction," from *The Irony of Regulatory Reform: The Deregulation of American Telecommunication* (New York: Oxford University Press, 1989), 1–44.

Hubel, David H., and Torsten N. Wiesel. "Receptive Fields, Binocular Interaction and Functional Architecture in the Cat's Visual Cortex." *Journal of Physiology 160*, no. 1 (1962): 106.

Istanbul Arastirmalari Enstitusu. "Anna L. Tsing—The Particular in the Planetary: Reimagining Cosmopolitanism Beyond the Human." YouTube Video. 1:33:30. April 8, 2021. https://www.youtube.com/watch?v=PqLiEo3baMc.

Jay, Megan. *The Defining Decade: Why Your Twenties Matter and How to Make the Most of Them Now* (New York: Twelve, 2012).

Kabali, Hilda K., Matilde M. Irigoyen, Rosemary Nunez-Davis, Jennifer G. Budacki, Sweta H. Mohanty, Kristin P. Leister, and Robert L. Bonner. "Exposure and Use of Mobile Media Devices by Young Children." *Pediatrics 136*, no. 6 (December 2015): 1044–50. DOI:10.1542/peds.2015-2151.

Kamp, David. *Sunny Days: The Children's Television Revolution That Changed America* (New York: Simon & Schuster, 2020).

Karlis, Nicole. "A Generation of Kids Has Used Social Media Their Whole Lives. Here's How It's Changing Them." *Salon* (February 2021). https://www.salon.com/2021/02/11/a-generation-of-kids-has-used-social-media-their-whole-lives-heres-how-its-changing-them/.

Kendall, Mikki. (@Karnythia). 2009. Twitter. https://twitter.com/Karnythia ?ref_src=twsrc%5Egoogle%7Ctwcamp%5Eserp%7Ctwgr%5Eauthor;.

Knight, Matthew. "'Miracle Material' Chips Away at Silicon Dominance." *CNN Business*. (October 2011). https://www.cnn.com/2011/10/12/tech /graphene-computer-chip-research/index.html.

Kriss, Alexander. *The Gaming Mind: A New Psychology of Videogames and the Power of Play* (New York: The Experiment, 2020).

Kunzru, Hari. "You Are Cyborg." *Wired* (February 1997). https://www.wired .com/1997/02/ffharaway/.

Langlois, Richard N. "'An Elephants' Graveyard': The Deregulation of American Industry in the Late Twentieth Century." University of Connecticut (July 2021). https://media.economics.uconn.edu/working/2021-11.pdf.

Larissa Buhler, Janina, Rebekka Weidmann, Jana Nikitin, and Alexandra Grob. "A Closer Look at Life Goals Across Adulthood: Applying a Developmental Perspective to Content, Dynamics, and Outcomes of Goal Importance and Goal Attainability." *European Journal of Personality 33*, no. 3 (May 2019): 359–84. DOI: 10.1002/per.2194.

Leonardo, Zeus, and Logan Manning. "White Historical Activity Theory: Toward a Critical Understanding of White Zones of Proximal Development." *Race and Ethnicity and Education* (November 2015). DOI: 10.1080/13613324.2015.1100988.

Leslie, Alan M. "Pretending and Believing: Issues in the Theory of ToMM." *Cognition 50* (1994): 211–38. DOI: 0010-277(93)00601-A.

Lewis, Danny. "Reagan and Gorbachev Agreed to Pause the Cold War in Case of an Alien Invasion." *Smithsonian Magazine* (November 2015). https://www .smithsonianmag.com/smart-news/reagan-and-gorbachev-agreed-pause -cold-war-case-alien-invasion-180957402/.

Lorde, Audre. "The Master's Tools Will Never Dismantle the Master's House." *Sister Outsider: Essays and Speeches* (Berkeley, CA: Crossing Press, 1984), 110–114. Print.

Mah, V. Kandice, and E. Lee Ford-Jones. "Spotlight on Middle Childhood: Rejuvenating the 'Forgotten Years.'" *Pediatrics Child Health 17*, no. 2 (February 2012): 81–3. DOI: 10.1093/pch.17.2.81.

Main, Thomas J. "Illiberalism in American Political Culture Today," from *The Rise of Illiberalism* (Washington D.C.: Brookings Institution Press, 2021).

"Make Way for Ducklings." *International Travel News*. https://www.intl travelnews.com/2010/09/go-east-young-duckshttps://www.intltravelnews .com/2010/09/go-east-young-ducks.

Maranto, Lauren. "Who Benefits from China's Cybersecurity Laws?" *Center for Strategic and International Studies* (June 2020). https://www.csis.org/blogs /new-perspectives-asia/who-benefits-chinas-cybersecurity-laws.

"Marketing Food to Children and Adolescents: A Review of Industry Expenditures, Activities, and Self-Regulation: A Federal Trade Commission Report to Congress." Federal Trade Commission. https://www.ftc.gov/reports /marketing-food-children-adolescents-review-industry-expenditures-activi ties-self-regulation.

Markus, Hazel Rose, and Alana Conner. *Clash!: How to Thrive in a Multicultural World* (New York: The Penguin Group, 2014).

Matz, S.C., M. Kosinski, G. Nave, and D.J. Stillwell. "Psychological Targeting As an Effective Approach to Digital Mass Persuasion." *PNAS 114*, no. 48 (November 2017): 12714–9. DOI: 10.1073/pnas/170966114.

McCaleb, Miriam, Patricia Champion, and Phillip J. Schluter. "Commentary: The Real (?) Effect of Smartphone Use on Parenting: A Commentary on Modecki et al. (2020)." *Journal of Child Psychology and Psychiatry 62*, no. 12 (2021): 1494–96. DOI: 10.1111/jcpp/13413.

McCarthy, John, Marvin L. Minsky, Nathaniel Rochester, and Claude E. Shannon. "A Proposal for the Dartmouth Summer Research Project on Artificial Intelligence, August 31, 1955." *AI magazine 27*, no. 4 (2006): 12.

McCloskey, Robert. *Make Way for Ducklings* (London: Penguin Young Readers Group, 1941).

Mcleod, Saul. "Konrad Lorenz's Imprinting Theory." *Simply Psychology*. 2018, updated 2021. https://www.simplypsychology.org/Konrad-Lorenz.html.

McNair, Carl. "Eyes on the Stars." *Storycorps*. https:storycorps.org/animation /eyes-on-the-stars/.

McNamee, Roger. "I Helped Create This Mess. Here's How to Fix It." January 28, 2019. https://time.com/magazine/us/5505429/january-28th-2019-vol -193-no-3-u-s/.

"Media Education. American Academy of Pediatrics. Committee on Public Education." *Pediatrics 104* no. 2, Part 1 (August 1999): 341–3. PMID: 10429023.

"Merrimack River at Risk." PBSNH Video. 56:14. July 14, 2020. https:// www.pbs.org/video/the-merrimack-river-at-risk-cq7o6d/.

"Michael Apted on the Incredible Journey of the 'Up' Series." *Film at Lincoln Center* (2019). https://www.filmlinc.org/daily/michael-apted-on-the -incredible-journey-of-the-up-series/.

Miranda, Lin-Manuel. "Hamilton: An American Musical," in *Hamilton: The Revolution*, edited by Jeremy McCarter. (New York: Grand Central Publishing, 2016).

Mishna, Faye, Michael Saini, and Steven Solomon. "Ongoing and Online: Children and Youth's Perceptions of Cyber Bullying." *Children and Youth Services Review 31*, no. 12 (2009): 1222–28.

Mitman, Gregg. "A Reflection of Plantationocene: A Conversation with Donna Haraway and Anna Tsing." *Edge Effects* (June 2019). https://edgeeffects.net /wp-content/uploads/2019/06/PlantationoceneReflections_Haraway _Tsing.pdf.

Mueller, Robert S., III. "Report on the Investigation into Russian Interference in the 2016 Presidential Election." US Department of Justice, Vol. I (March 2019). https://www.justice.gov/archives/sco/file/1373816/download.

Mullaney, Thomas S., Benjamin Peters, Mar Hicks, and Kavita Philip, editors. *Your Computer Is on Fire* (Cambridge, MA: The MIT Press, 2021).

Musgrave, Shawn. "How White Nationalists Fooled the Media about Florida Shooter." Politico.com. February 2018. https://www.politico.com /story/2018/02/16/florida-shooting-white-nationalists-415672.

National Association of Education for Young Children. Technology and Interactive Media as Tools in Early Childhood Programs Serving Children from Birth through Age 8. 2012. https://www.naeyc.org/sites/default/files /globally-shared/downloads/PDFs/resources/position-statements/ps_tech nology.pdf.

Nelson, Garrison. "The Rise of the 'Straight, White, Angry Male' Voter." Paper presented at Broken: Barriers, Parties, and Conventional Wisdom in 2016: American Elections Academic Symposium. Saint Anselm College. March 17–18, 2017. https://www.anselm.edu/sites/default/files/Documents/NHIOP /Symposium%20Program%202017.pdf.

Nesse, Randolph M. "The Evolution of Hope and Despair." *Social Research 66*, no. 2 (Summer 1999): 429–69. http://www.jstor.org/stable/40971332.

Newcomer, Eric, and Brad Stone. "The Fall of Travis Kalanick Was a Lot Weirder and Darker Than You Thought." *Bloomberg Businessweek* (January 2018). https://www.bloomberg.com/news/features/2018-01-18/the-fall-of -travis-kalanick-was-a-lot-weirder-and-darker-than-you-though.

Nugent, J. Kevin, Constance H. Keefer, Susan Minear, Lise C. Johnson, and Yvette Blanchard. *Understanding Newborn Behavior and Early Relationships: The Newborn Behavioral Observations (NBO) System Handbook* (Baltimore: Paul H. Brookes, 2007).

Oduro, Serena. "A New AI Lexicon: Black Women Best." *Medium* (September 2021). https://medium.com/a-new-ai-lexicon/a-new-ai-lexicon-black -women-best-14eb5d059b2a.

Oduro, Serena Dokuaa. "Do We Need AI or Do We Need Black Feminism? A Poetic Guide," from *Fake AI* (Manchester, UK: Meatspace Press, 2021), 89–95. https://ia801504.us.archive.org/6/items/fake-ai/Fake_AI.pdf.

Pape, Robert A., and Keven Ruby. "The Face of American Insurrection: Right Wing Organizations Evolving into a Violent Mass Movement." *Chicago Project on Security and Threats*. The University of Chicago. January 28, 2021. https://d3qi0qp55mx5f5.cloudfront.net/cpost/i/docs/americas_insur rectionists_online_2021_01_29.pdf?mtime=1611966204.

Parida, Tulsi, and Aparna Ashok. "Consolidating Power in the Name of Progress: Technosolutionism and Farmer Protests in India," from *Fake AI* (Manchester,

UK: Meatspace Press, 2021), 164. https://ia801504.us.archive.org/6/items/fake -ai/Fake_AI.pdf.

Patton, Desmond Upton. Interview by Loretta L.C. Brady, December 2021.

Perrin, Andrew. "5 Facts about Americans and Video Games." Pew Research Center (September 2018). https://www.pewresearch.org/fact-tank/2018/09/17/5 -facts-about-americans-and-video-games/.

Perry, Tam E., Luke Hassevoort, Nicole Ruggiano, and Natali Shtompel. "Applying Erikson's Wisdom to Self-Management Practices of Older Adults: Finding from Two Field Studies." *Research on Aging 37*, no. 3 (April 2015): 253–74. DOI: 10.1177/016402751427974.

Poole, Mary Elizabeth. "Securing Race and Ensuring Dependence: The Social Security Act of 1935" (Rutgers, The State University of New Jersey–New Brunswick, 2000).

"President's Committee of Advisors on Science and Technology." *Panel on Educational Technology* (March 1997). https://clintonwhitehouse3.archives .gov/WH/EOP/OSTP/NSTC/PCAST/k-12ed.html.

Prilleltensky, Isaac. "Mattering at the Intersection of Psychology, Philosophy, and Politics." *American Journal of Community Psychology 65*, no. 1–2 (2019): 16–34. DOI: 10.1002/ajcp.12368.

Prince, Dana. "What About Place? Considering the Role of Physical Environment on Youth Imagining of Future Possible Selves." *Journal of Youth Studies* 17, no. 6 (2014): 697–716.e.

Protecting Kids Online: Snapchat, TikTok, and YouTube. (October 2021). Written testimony of Leslie Miller, vice president of YouTube government affairs and public policy. https://www.commerce.senate.gov/services/files/2FBF8DE5 -9C3F-4974-87EE-01CB2D262EEA.

Przybylski, Andrew K., and Netta Weinstein. "A Large-Scale Test of the Goldilocks Hypothesis: Quantifying the Relationship Between Digital Screen Use and the Mental Well-being of Adolescents." *Psychological Science 28*, no. 2 (2017): 204–7. DOI: 10.1177/0956797616678438.

Pyne, Irene. "YouTubers Under 18 Who Are Earning Up to US $22 million per Year: Ryan ToysReview, Boram Tube Vlog and More." *Style.* October 17, 2019. https://www.scmp.com/magazines/style/news-trends /article/3033320/ryan-toysreview-boram-tube-vlog-4-youtubers-under-18.

Rooks, Belvie, Kabir Hypolite, Loretta Brady, and Bithiah Carter. "Panel 6, I'll Fly Away: How We Move Forward from Here. In Honor of Anya Dillard." Virtual Panel. Crossing River Jordan: Healing Racial Wounds Through Accountability and Truth Telling. October 23, 2021.

Rose, Megan. "The Average Parent Shares Almost 1,500 Images of Their Child Online Before Their 5th Birthday." *Parent Zone.* https://parentzone org.uk/article/average-parent-shares-almost-1500-images-their-child-on line-their-5th-birthday.

Rothstein, Richard. *The Color of Law: A Forgotten History of How Our Government Segregated* (New York: Liveright Publishing, 2017).

Rotter, Julian B. "Social Learning Theory." *Expectations and Actions: Expectancy-Value Models in Psychology 395* (1982).

Scheck, Justin, Newley Purnell, and Jeff Horwitz. "The Facebook Files: Facebook's Staff Flags Criminals, But Company Often Fails to Act—Documents Show Alarm About What Is on the Site Outside U.S., Where User Numbers Are Huge." *Wall Street Journal* (September 2021). https://www.proquest.com/docview/2573296131?accountid=13640;.

Sherman, Lauren E., Leanna M. Hernandez, Patricia M. Greenfield, and Mirella Dapretto. "What the Brain 'Likes': Neural Correlates of Providing Feedback on Social Media." *Social Cognitive and Affective Neuroscience* 13, no. 7 (July 2018): 699–707. DOI: 10.1093/scan/nsy051.

Smith, Leslie. "Piaget's Infancy Journal: Epistemological Issues." *Constructivist Foundations 14*, no. 1 (2018): 85–7.

Smith, Yves. "Michael Hudson: On Debt Parasites." [Podcast]. *Naked Capitalism* (January 2022). https://www.nakedcapitalism.com/2022/01/michael-hudson-on-debt-parasites.html.

Smogorzewska, Joanna, Grzegorz Szumski, and Pawel Grygiel. "Theory of Mind Goes to School: Does Educational Environmental Influence the Development of Theory of Mind in Middle Childhood?" *PLoS ONE 15*, no. 8 (August 2020). DOI:10.1371/journal.pone.0237524.

Solomon, Feliz. "What 'The Hunger Games' Three-Finger Salute Means to Protesters Across Asia." https://www.wsj.com/articles/what-the-hunger-games-three-finger-salute-means-to-protesters-across-asia-11613649604, accessed February 2021.

Straight, Andrew. "Why Automated Content Moderation Won't Save Us," from *Fake AI*. (Manchester, UK: Meatspace Press, 2021), 148–58. https://ia801504.us.archive.org/6/items/fake-ai/Fake_AI.pdf.

Suderman, Alan, and Joshua Goodman. "Amid the Capitol Riot, Facebook Faced Its Own Insurrection." *AP NEWS* (October 2021). https://apnews.com/article/donald-trump-technology-business-social-media-media-07124025bdbeba98a7c7b181562c3c1a.

Takeuchi, Lori, and Reed Stevens. "The New Coviewing: Designing for Learning through Joint Media Engagement." December 8, 2011. Report of the Jean Gantz Cooney Center. https://joanganzcooneycenter.org/publication/the-new-coviewing-designing-for-learning-through-joint-media-engagement/.

TEDxTalks. "They Are Children: How Posts on Social Media Lead to Gang Violence." YouTube Video. 12:19. May 15, 2017. https://www.youtube.com/watch?v=BmlvOGh7Spo.

Thakor, Mitali. "Capture Is Pleasure," from *Your Computer Is on Fire*. Ed. Thomas S. Mullaney, Benjamin Peters, Mar Hicks, and Kavita Philip (Cambridge, MA: The MIT Press, 2021).

Thomas, Balmès, director. *Babies*. Universal Studios Home Entertainment, 2010. 1:19:00.

Tronick, Ed, and Marjorie Beeghly. "Infants' Meaning-Making and the Development of Mental Health Problem." *American Psychologist 66*, no. 2 (Winter 2011): 107–19. DOI: 10.1037/a0021631.

Tsing, Anna. "Arts of Inclusion, or How to Love a Mushroom." *Mānoa 22*, no. 2 (Winter 2010): 191–203. https://www.jstor.org/stable/41479491.

Uber Team. "Uber's New CEO." Uber Newsroom (August 2017). https://www.uber.com/newsroom/ubers-new-ceo-3/.

Ullman, Ellen. *Close to the Machine: Technophilia and Its Discontents* (New York: Picadorusa, 2012).

Ulman, Ellen. *Life in Code: A Personal History of Technology* (New York: MCD, 2017).

"Understanding the Development of Attachments Bonds and Attachment Behavior Over the Life Course," from Daniel P. Brown and David S. Elliott, editors, *Attachment Disturbances in Adults: Treatment for Comprehensive Repair* (New York: W. W. Norton, 2016), 75–102.

UN News Staff. "Urgent Action Needed over Artificial Intelligence Risks to Human Rights." *UN News*. September 21, 2021. https://news.un.org/en/story/2021/09/1099972 .

"Urie Bronfenbrenner." College of Human Ecology. https://bctr.cornell.edu/about-us/urie-bronfenbrenner.

Vaala, Sarah, Anna Ly, and Michael H. Levine. *Getting a Read on the App Stores: A Market Scan and Analysis of Children's Literacy Apps* (New York: The Joan Ganz Cooney Center at Sesame Workshop, Fall 2015), 1–50. https://www.joanganzcooneycenter.org/wp-content/uploads/2015/12/jgcc_getting aread.pdf.

Villegas, Seth. Interview by Loretta L.C. Brady. Zoom interview. Manchester, New Hampshire. October 2021.

Vygotsky, Lev S. "Play and Its Role in the Mental Development of the Child." *Soviet Psychology 5*, no. 3 (1967): 6–18.

Warner, Judith. *Perfect Madness: Motherhood in the Age of Anxiety* (New York: Riverhead Books, 2006).

Wartella, Ellen. "A Brief History of Children and Media Research. Workshop on Media Exposure and Early Childhood Development." Presented at Northwestern University. January 25, 2018. https://www.nichd.nih.gov/sites/default/files/2018-03/WartellaHistChildMediaResearch.pdf, accessed September 19, 2021.

"Watch: House Hearing on Social Media Reform, with Facebook Whistle-blower Frances Haugen." *PBS News Hour* (December 2021). https://www.pbs.org/newshour/economy/watch-live-house-hearing-on-social-media-reforms-with-facebook-whistleblower-frances-haugen.

Webster-Stratton, Carolyn. "Cross-Cultural Collaboration to Deliver the Incredible Years Parent Program." *Incredible Years* (2006), 1–34. https://incredibleyears.com/wp-content/uploads/training-interpreters-deliver-cross-cultural-colaboration_06.pdf.

Wellman, Henry M., and Karen Lind. *Reading Minds: How Childhood Teaches Us to Understand People* (New York: Oxford University Press, 2020).

Wells, Georgia, and Jeff Horwitz. "The Facebook Files: Facebook's Effort to Attract Preteens Reaches Back Years—Documents Shows Moves Came in Response to Competition from Snapchat, TikTok." *Wall Street Journal* (September 2021). https://www.proquest.com/docview/2577278884?accountid=13640;.

Wells, Georgia, Jeff Horwitz, and Deepa Seetharaman. "The Facebook Files: Facebook Knows Instagram Is Toxic for Teen Girls, Its Research Shows—Internal Document Shows a Youth Mental-Health Issue That Facebook Plays Down in Public." *Wall Street Journal.* (September 2021). https://www.proquest.com/docview/2572410901?accountid=13640&parentSessionId=YW4qkaAdJbrpLHiL5TY20t%2BAJKN7xytSr0ayv70%2BKMg%3D;.

Winnicott, D.W. "Transitional Objects and Transitional Phenomena," from *Playing & Reality* (London: Tavistock Publications, 1971), 1–18.

Xavier, Jean, Julien Magnat, Alain Sherman, Soizic Gauthier, David Cohen, and Laurence Chaby. "A Developmental and Clinical Perspective of Rhythmic Interpersonal Coordination: From Mimicry Towards the Interconnection of Minds." *Journal of Physiology-Paris 110*, 4-part B (November 2016): 420–6. DOI: 10.1016/j.physparis.2017.06.001.

Younger, Shannan. "Info Parents Need to Know About the Blue Whale Challenge." *Chicago Parent.* July 2017. https://www.chicagoparent.com/parenting/info-parents-need-to-know-about-the-blue-whale-challenge/.

Mullaney, Thomas S., Benjamin Peters, Mar Hicks, and Kavita Philip, editors. *Your Computer Is on Fire.* (Cambridge, MA: The MIT Press, 2021).

Zialcita, Paolo. "Facebook Pays $643,000 Fine for Fine in Cambridge Analytica Scandal." NPR. October 2019. https://www.npr.org/2019/10/30/774749376/facebook-pays-643-000-fine-for-role-in-cambridge-analytica-scandal.

Index

About the Author and Contributors

Loretta L.C. Brady, PhD, is a licensed clinical psychologist and professor of psychology at Saint Anselm College, where she directs the Community Resilience and Social Equity Lab (CRSEL). She previously served as codirector for the Center for Teaching Excellence. She serves on the boards of several task forces and local nonprofits, including youth-serving organizations and healthcare systems. She is the author of *Bad Ass & Bold: A transformative approach to planning with your loves, dreams, and realities in mind* (www.badassandbold.com). Her award-winning writing has been recognized by the New England Society of Children's Book Writers & Illustrators, Jack Jones Literary Arts, and the New England Press Association for her work in children's picture books, young adult novels, creative nonfiction, and advice. Her work has appeared in *New Hampshire Business Review* and *Business NH Magazine*, and she has been a source for the *New York Times*, *USA Today*, and the *Washington Post* on issues related to workforce development and resilience. Her career includes awards for her statewide service and social-equity efforts, a TEDx Talk (2014, *Bootstraps Are Bullsh*t*), a Fulbright fellowship (Cyprus 2013), a McNair fellowship (UNH 1998), and recognition as a Top Ten College Woman (*Glamour* 1998). She lives in Manchester, New Hampshire, with her family and socially distancing dog, Zelda.

CONTRIBUTORS

Michael Alcée, PhD, is a clinical psychologist in private practice in Tarrytown, New York, and a mental health educator at the Manhattan

School of Music. Specializing in the psychology of artists, everyday creativity, and the professional development of therapists, he is the author of *Therapeutic Improvisation: How to Stop Winging It and Own It as a Therapist.* His contributions have appeared in the *Chicago Tribune,* the *New York Times,* NPR, the *New York Post,* Salon.com, and on the TEDx stage.

Erica DeMatos is a junior at Lesley University in Cambridge, Massachusetts, where she is earning a degree in English. Her literary interests include nonfiction essays and memoirs, and she intends to pursue a career in editing. In her free time, Erica likes to go to the Boston Public Garden to people watch, read, and write.

Brenna Leach received her BA in psychology from Saint Anselm College in 2022, where she met and worked closely with Dr. Loretta L.C. Brady on *Technology Touchpoints.* She has worked closely within the Manchester youth community, especially with individuals facing adversity, helping to develop and foster mentoring programs. She lives in Massachusetts with her family, but she has fallen in love with the natural beauty and atmosphere of New Hampshire.

CPSIA information can be obtained
at www.ICGtesting.com
Printed in the USA
BVHW040840100722
640702BV00005B/2

9 781538 163924